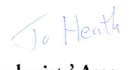

Geologists' Asso.

THE GEOLOGY OF BARCELONA: AN URBAN EXCURSION GUIDE

Wes Gibbons and Teresa Moreno

Guide Series Editor: Susan B. Marriott

CONTENTS
Page

i

Geology of Barcelona

ACKNOWLEDGEMENTS

A guide such as this involves the condensation and overview of detailed research work published by many geologists over the last 200 years or so. We have done our best to illustrate our recognition of this fact by providing a long reference list of publications in Catalan, Spanish, French and English for further reading and research, and we apologise to anyone we have inadvertently omitted to mention. The guide is also an attempt to convey enthusiasm for geology and how an informed appreciation of the rocks and landscape can so much enhance enjoyment of a visit even to a big city like Barcelona. In this context we would especially like to thank Greg Power who carefully read and commented on the text. His unwavering support and geological enthusiasm, undiminished over the 40 years during which WG has known him, are truly appreciated. Thanks also to David Maynard who helped with the literature research, and to our excellent Series Editor Susan Marriott for her detailed and efficient work on the manuscript and figures.

Wes Gibbons and Teresa Moreno

Geology of Barcelona

PREFACE

The purpose of this guide is to describe the geology and scenery of the Mediterranean city of Barcelona. Although a popular tourist destination much in demand for its cultural and historical riches, Barcelona is equally interesting from a geological viewpoint. The three excursions described in this guide can easily be completed in a short citybreak visit using public transport, ideally during weekdays to avoid weekend crowds, and involving a maximum of 4 nights in a hotel (3 excursion days plus arrival and departure days). The excursions visit many of the scenically most attractive parts of the city, including places as yet virtually undiscovered by tourists, and provide an opportunity to see Barcelona from a different perspective.

The most convenient location to use as a base in the city is basically anywhere within easy reach of either the green metro line L3 or the underground FGC lines running from Plaça Catalunya to Sabadell (line S2) and Terrassa (line S1). Thus anywhere in the Medieval city centre close to Las Ramblas and Plaça Catalunya, or the more bohemian district of Gràcia (near FGC station Gràcia and metro station Fontana) would be perfect, although other areas along metro L3, such as Les Corts (which is away from the tourist centre and therefore can be less expensive) are also convenient choices. For those arriving at the airport a taxi costing around 30 euros (2011) is the most rapid and direct way to reach a hotel, although if flying in with only hand baggage then a multiperson T10 ticket bought at the Terminal 2 airport train station will currently (2011) get you into the city and on the underground rail system for less than 1 euro. Once the long-awaited new L9 metro line opens, this will provide a simple, direct and inexpensive connection between the airport and city. Use of public transport in Barcelona is safe but inevitably requires some degree of vigilance and awareness (but no paranoia): theft of wallets and bags from unsuspecting tourists is at least as common as in any other European city.

No special field equipment is required other than comfortable, protective footwear, a compass, hand lens and a street map. If the weather forecast predicts rain, then waterproofs and umbrella are obviously advisable, although a hat, cream and water to avoid sunburn and dehydration are more usually *de rigueur* in Barcelona. Use of a hammer is not appropriate: part of the route is in protected parkland (where damage to exposures is prohibited) and the rest in the streets of the city itself (where a hammer is an offensive weapon). While maps in addition to those presented in this guide are not essential to completing the excursions, a good range of geological maps can be obtained from the Institut Cartogràfic de Catalunya in Montjuïc Park (see their website for the maps on offer and location of mapshop: http://www.icc.cat/eng/Home-ICC/Home). Tourist information can be obtained by visiting the municipal website http://www.bcn.es/english/ihome. htm. Finally, suitable hotels can be chosen via large search engine websites such as tripadvisor.com and booking.com where recent opinions from fellow travellers are freely available.

Geology of Barcelona

LIST OF FIGURES

28. Use of Miocene Montjuïc Formation sandstone in Roman funeral monuments.
29. The large roseton window above the front entrance to the earthquake damaged church of Santa Maria del Mar (Excursion 3, Location 14).

LIST OF TABLES

Geology of Barcelona

INTRODUCTION

Barcelona is one of Europe's most geologically interesting cities. Although examining rocks is not exactly a priority for visiting tourists, many of the attractions they come to enjoy owe much to a varied geomorphology developed over the complex geology which characterises this part of Catalonia. The geology has shaped the mountainous backdrop to the city, the hilltop urban parks, the accessible coastline and consequent long history of human occupation. It has clearly influenced the incorporation of natural landforms into Gaudí's architecture, as any geologically minded visitor to Park Güell or Montserrat will readily appreciate. This geology includes Variscan basement and post-Variscan cover within an Alpine tectonic setting, so that within the city limits one can examine igneous, metamorphic and sedimentary lithologies. All three Phanerozoic eras are represented by exposures of Cambro-Ordovician to Permian, Triassic and Neogene rocks.

The variety offered by Barcelona's geology reflects the position of the city on the flanks of the Catalan Coastal Ranges (CCR), an Iberian intraplate mountain belt that runs parallel to the Mediterranean coast and separates the Valencia Trough from the Ebro **foreland basin**, which is in turn bordered to the north by the Pyrenees (Fig. 1). The CCR comprise a "Basin and Range"-style double chain of low mountains cored by Variscan basement uplifted during Palaeogene compression and later modified by Neogene extensional collapse during the opening of the Valencia Trough. Thus, in the city there are both Palaeogene thrusts directed NW towards the Ebro Basin, and Neogene extensional faults dipping SE beneath the Mediterranean Sea, with the whole urban area lying within a major fault system bordering the Barcelona Basin. Another important structural control on geomorphology is provided by NW-SE striking fault systems such as that traceable SE from Montserrat to the Llobregat River delta and out into the Mediterranean where it is expressed as an undersea volcanic transverse **fracture zone** lying north of Mallorca (Fig. 1).

The tectonic setting outlined above and illustrated by Figure 1 is given more detail in Figure 2, which focuses on the onshore geology of the Barcelona area. The oldest (Palaeozoic) rocks are abundantly exposed in the Collserola Hills which rise immediately behind the city to a height of over 500 m (Fig. 3). These basement rocks are mostly **metasediments** and range from Cambro-Ordovician to Carboniferous in age. They are intruded by late Variscan post-orogenic **batholithic** rocks (mostly **granodiorites**) of the Montnegre Massif which extends NE to meet the sea at the Costa Brava (Fig. 2). Together, the Montnegre and Collserola rocks form part of the seaward chain or "Littoral Range" of the CCR, and are locally overlain unconformably by Triassic sediments. The Littoral Range continues SW across the Llobregat river valley to form the Garraf Massif where the Palaeozoic basement disappears westwards beneath Mesozoic cover (Fig. 2). The Garraf Massif is notable for preserving not only Triassic rocks but also a Jurassic-

Introduction

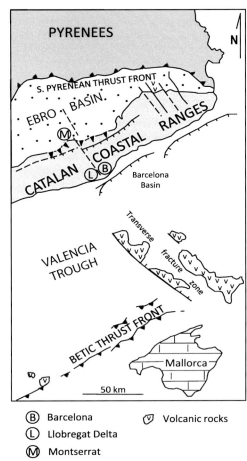

Figure 1. The geotectonic setting of Barcelona. The city lies on the Mediterranean coast on the SE side of the Catalan Coastal Ranges (CCR) which separate the Ebro (Pyrenean) foreland basin from the submerged Valencia Trough. Whereas the main structural controls in the CCR (thrusts and extensional faults) run NE-SW, the area is also crosscut by NW-SE transverse fault zones. These transverse structures (such as that running through the Llobregat Delta) have locally influenced patterns of Cenozoic tectonics and sedimentation (as at Montserrat) and provided a focus for both subaerial and underwater Cenozoic basaltic volcanism.

Figure 2. (Facing page) Map outlining the geology of the Barcelona area. The city is built mainly on Quaternary deposits between the Besòs and Llobregat deltas, and is bordered to the NW by Palaeozoic basement (metasediments and granitoid plutons) of the Littoral Range. A thick Cenozoic sedimentary infill lies between the Littoral and Pre-Littoral ranges and was deposited during major SE-directed extension following earlier Palaeogene syn-sedimentary NW-directed thrusting over sediments of the Ebro Basin. The cross section emphasises the listric normal fault system producing the SE-directed extension. Barcelona lies perched on the faults splaying out from that controlling the Barcelona Basin. To the SW there are extensive outcrops of Mesozoic sediments. For further detail see Roca *et al.* (1999), Maillard & Mauffret (1999), Gaspar-Escribano *et al.* (2002) and ICC (2005). (PLR: Pre-Littoral Range; MF: Morrot Fault.)

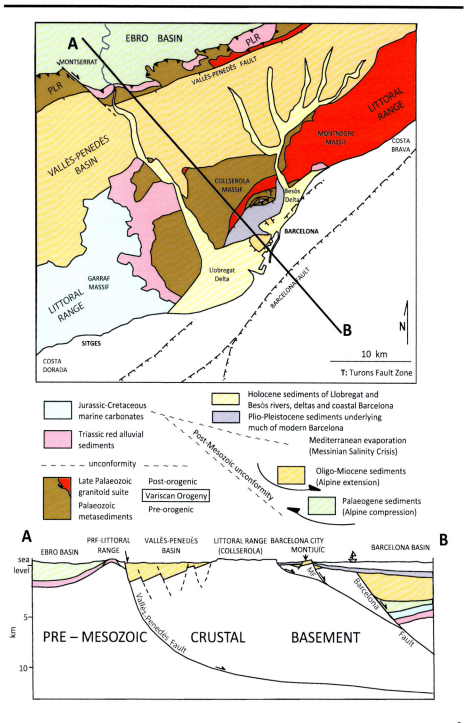

Jurassic-Cretaceous marine carbonates

Triassic red alluvial sediments

~ ~ ~ ~ ~ unconformity ~ ~ ~ ~ ~

Late Palaeozoic granitoid suite

Palaeozoic metasediments

Post-orogenic
Variscan Orogeny
Pre-orogenic

Holocene sediments of Llobregat and Besòs rivers, deltas and coastal Barcelona

Plio-Pleistocene sediments underlying much of modern Barcelona

Mediterranean evaporation (Messinian Salinity Crisis)

Oligo-Miocene sediments (Alpine extension)

Palaeogene sediments (Alpine compression)

Post-Mesozoic unconformity

3

Introduction

Cretaceous marine carbonate sequence not seen elsewhere in the Barcelona area (Fig. 2). There is therefore a marked contrast in coastal geology and scenery on either side of Barcelona. To the northeast, beyond the River Besòs delta, are the **granitoid** cliffs of the Costa Brava, whereas to the southwest, along the Sitges coast beyond the River Llobregat delta, Mesozoic limestones of the Costa Dorada crop out (Fig. 2).

Behind the Littoral Range to the NW is a prominent linear basin (Vallès-Penedès) which lies in the **hanging wall** of a major SE-dipping normal fault (Vallès-Penedès Fault) and contains a mostly mid-late Cenozoic sedimentary infill (Fig. 2). Prior to the extensional movements along structures such as the Vallès-Penedès Fault, the area had been thrust NW over early Cenozoic sediments in the Ebro Basin to produce a Palaeogene mountain range. It was the extensional collapse of these mountains that resulted in the modern topography of the CCR, with the Vallès sedimentary infill separating the Littoral and Pre-Littoral ranges. The present-day structure of the area is thus dominated by major, SE-dipping listric faults, which curve to crustal depths >10 km and have a long history of polyphase movements during alternating periods of extension (Mesozoic and mid-late Cenozoic) and compression (early Cenozoic). The dominant structure, larger even than the Vallès-Penedès Fault, is the Barcelona Fault, which (during late Oligocene-Miocene times) has been estimated to have extended the crust by around 11 km (Gaspar-Escribano *et al.*, 2004). Minor splays northwestward from this major fault run on land beneath Barcelona and influence the topography of the city area. One such splay, the Morrot Fault, is responsible for the prominent hill of Montjuïc (Fig. 2), the Miocene sediments of which rise above the city from the south, producing the site chosen to stage the 1992 Olympic Games. Another important fracture zone links the Barcelona Fault with the Littoral Range and has resulted in a distinctive chain of isolated, fault-bounded hills (Les Turons) rising abruptly from a low-lying littoral plain (Pla de Barcelona) underlain by Plio-Pleistocene sediments.

Visits to the Collserola Hills of the Littoral Range, Les Turons Fault Zone, Montjuïc Hill, and the Plio-Pleistocene coastal plain are all included within the three excursion routes described below, following an overview of the stratigraphic history of the area.

4

Figure 3. The Collserola Hills form a scenic backdrop to Barcelona and rise to a maximum height of 512 m at Tibidabo. These hills belong within the Littoral Range of the Catalan Coastal Ranges and expose Palaeozoic basement (mostly metasediments and granodiorite). The route of Excursion 1 described in this guide follows these hills (from extreme left to right in the photo), using a track (Passeig de les Aigües) which contours midway up the hillside. View looking NW from Montjuïc across Plaça d'Espanya.

Previous publications

PREVIOUS PUBLICATIONS

One of the earliest studies of Catalonian geology was made by Charles Lyell during a tour of France and Spain in the summer of 1830, a time when he was writing and publishing his three-volume Principles of Geology (Lyell, 1830, 1832, 1833; Virgili, 2007a). Lyell's journey included a visit to Barcelona, where he showed a particular interest in collecting fossils from the marine sediments of Montjuïc (Virgili, 2007b), rocks which would later be recognised as belonging to the Miocene epoch (Lyell was to introduce the term Miocene in volume three of his trilogy). It was however the later work of other 19th century geologists such as Vézian (1856) and, especially, Almera (1891–1900) that produced the first regional cartographic studies and established the geographic distribution of the main lithologies cropping out in the Barcelona area. The series of geological maps of the Barcelona area produced in the late 19th century by Jaume Almera and others in turn aided more detailed applied studies, particularly those concerned with elucidating the hydrogeological network of potable waters beneath the city (e.g. Rubio & Kindelán, 1910). The contaminated nature of many wells had led to frequent outbreaks of typhoid, one of the worst of which killed over 2,200 people in 1914–15 (Conillera, 1991; Stallybrass, 1931).

During the 20th century a number of research papers further elucidated the geology of the Barcelona region, although until the last decade these were mostly written in Catalan or Spanish, restricting their distribution outside Iberia. Among the most notable of these works published on the pre-Neogene geology (but not written in English) are those by San Miguel (1929) on the metamorphic basement around Tibidabo; Marcet (1933, 1960) on Palaeozoic stratigraphy; Depape and Solé Sabaris (1934) on Triassic in the Montgat area; Llopis *et al.* (1969) on the Silurian-Devonian boundary; Batlle (1976) on the geotechnics of the Palaeozoic subcrop; Cabrera and Satanach (1979) and Valenciano and Sanz (1981) on Palaeozoic and Triassic outcrops in NE Barcelona; Anadón *et al.* (1979, 1983) on tectonostratigraphy of the Catalan Coastal Ranges (CCR); Enrique (1979) on Variscan granitoid intrusions; Julivert *et al.* (1985) on Ordovician-Silurian graptolites; Solé de Porta *et al.* (1987) on the Triassic; Gil Ibarguchi and Julivert (1988) on the metamorphic geology of the Collserola Hills and García-López *et al.* (1990) on Silurian-Devonian palaeostratigraphy. Research contributions published in English on the Palaeozoic geology of the area include those by Villas *et al.* (1987) on Ordovician palaeostratigraphy; Julivert and Durán (1990a, b) on Palaeozoic structure and stratigraphy; Gil Ibarguchi *et al.* (1990) on Lower Palaeozoic igneous rocks; Sebastian *et al.* (1990) on Variscan metamorphism; Masana (1996) on neotectonics; Solé *et al.* (2002) on Variscan granitoid geochronology and Plusquellec *et al.* (2006) on Devonian and Mississippian corals.

With regard to the younger strata exposed in the Barcelona area, Lyell was not the only researcher to focus attention on the geology of Montjuïc (see the extensive bibliography compiled by Vía Boada & Padreny, 1972), with detailed studies published on this Miocene inlier spanning the early 19th to 21st centuries (from Yáñez, 1822 to Gómez-Gras et al., 2001). Overlying the Miocene succession are Pliocene rocks which, although generally not well exposed, were nevertheless recognised by Almera in the late 19th century and subsequently further studied by Elías (1931), Gillet and Vicente (1961), Valenciano and Sanz (1967), and Alborch et al. (1980). Finally, the extensive Quaternary deposits which immediately underlie much of the city of Barcelona were studied in detail by Llopis (1942), Marqués (1966, 1975), Julià (1977), Checa et al. (1988), Sanz Parera (1988) and have been the subject of many modern studies (see below).

Pioneering land-based studies synthesising the complex tectonostratigraphic evolution of the CCR (e.g. Anadón et al., 1979) were supplemented in the 1980s by the arrival of new data emerging from petroleum exploration in the Gulf of Valencia region. These data helped stimulate a proliferation of research publications on NW Mediterranean tectonics and sedimentation, mostly written in English and in internationally available journals and books (Dañobeitia et al., 1990; Fontboté et al., 1990; Clavell & Berastegui, 1991; Banda & Santanach, 1992; Bartrina et al., 1992; Roca & Desegaulx, 1992; Roca & Guimerà, 1992; Salas & Casas, 1993; Cabrera, 1994; Sàbat et al., 1995; Anadón & Roca, 1996; Banda, 1996; Cabrera & Calvet, 1996; Friend & Dabrio, 1996; Martinez del Olmo, 1996; Torné, 1996; Maillard & Maufrett, 1999; Roca et al., 1999; Gaspar-Escribano et al., 2001, 2002, 2004; Amblàs et al., 2006). Related to this improved understanding of the broader geotectonic setting has been a particular focus on the interplay between Alpine tectonics and sedimentation along the SE margin of the Ebro Basin (López-Blanco et al., 2000; Barberá et al., 2001; López-Blanco, 2002) and a renewed interest in the history of seismic activity in Catalonia (Olivera et al., 1991, 1992; ICC, 1999; Lunar et al., 2002). Much of the work on the geology of the CCR published prior to 2002 has been synthesised and placed within a regional context within various chapters of the Geological Society of London book overviewing the geology of Spain (Gibbons & Moreno, 2002).

There has also been much recent interest in the depositional history of Plio-Pleisto-Holocene sediments along the Catalan coastal margin, combining recent work on NW Mediterranean marine geology with modern interest in climate change (Díaz et al., 1990; Ercilla et al., 1994; Jalut et al., 2000; Gámez et al., 2005, 2009; Simó et al., 2005; Lafuerza et al., 2005). A detailed summary and review of the geology of the seaward side of Barcelona city (between the Besòs and Llobregat deltas) has recently been published in Catalan by Riba and Colombo (2009). Improved understanding of Quaternary stratigraphy beneath the city has once again aided hydrological studies, such as in the Llobregat delta which has long been a critical source of water for Barcelona city (e.g. Vázquez-Suñé et al., 2006). Finally, an important stimulus for improved understanding of the urban geology of Barcelona has been provided by the needs of major 21st century engineering works such as the ongoing construction of new underground train lines (notably the Spain-France high speed rail-link and the L9 which, at over 42km,

Previous publications

will be the longest metro line in Europe), with the resulting recent and ongoing production of new maps (e.g. ICC, 2000, 2002, 2005) including large scale urban geology maps (IGC/ICC, 2009a, b).

	NEOGENE	ALPINE OROGENY	Deposition of Montjuïc sandstones during extensional collapse
	PALAEOGENE		Compressional uplift of Pyrenees and Catalonian Coastal Ranges Iberian Plate collision with Europe
100 Ma	CRETACEOUS		
			Long period of widespread marine carbonate deposition on Iberian Plate rotating between Africa and Europe
	JURASSIC		
200 Ma			
	TRIASSIC		"Buntsandstein" Facies fluvial red bed deposition
			Prominent unconformity
	PERMIAN	VARISCAN OROGENY	Granodioritic batholith intrusion and contact metamorphism
300 Ma	CARBONIFEROUS		
			"Culm Facies" sandstones
400 Ma	DEVONIAN		Pelagic limestones and shales of Olorda Fm.
	SILURIAN		Pelagic limestones of La Creu Fm. Pelagic black shales and cherts
	ORDOVICIAN		Low-grade metamorphic rocks of the Collserola Massif: mostly siliciclastic metasediments
500 Ma	CAMBRIAN		

Table 1. Stratigraphic column listing the pre-Quaternary rocks exposed in the Barcelona area and summarising relevant geological events.

STRATIGRAPHIC AND PALAEOGEOGRAPHIC OVERVIEW

The oldest rocks in the Catalan Coastal Ranges (CCR) are unfossiliferous, strongly **foliated** metasediments thought to be of Cambrian to Lower Ordovician age (Table 1). Away from proximity to late-Variscan plutonic intrusions, the **metamorphic grade** of these rocks is low (lower **greenschist facies**), and as a consequence they have commonly been referred to informally as the "sericitic slates", which gives a good idea of the typical lithology encountered in the field. This mostly siliciclastic sedimentary sequence was deposited in cold water at high southern latitudes (>70°S) on the northern margin of the African part of the Gondwanan continent (Robardet & Gutiérrez-Marco, 2002). Although specific **lithostratigraphic correlations** are not possible, given the transposed bedding and the lack of palaeostratigraphic data, these rocks have been broadly correlated with similar sequences in the Cantabrian Zone, the Pyrenees and the Montagne Noire (Ábalos *et al.*, 2002; Gutiérrez-Marco *et al.*, 2002; Liñán *et al.*, 2002). Above the Cambro-Lower Ordovician metasediments, the Palaeozoic succession continues with Upper Ordovician, Silurian and Devonian sediments (Table 1) yielding fossils and therefore with better stratigraphic control. The Palaeozoic basement rocks exposed in the hills behind Barcelona are mostly **metapelites** and **metapsammites**, with intercalations of **metaquartzites**, **metalimestones**, **calc-silicate** rocks and **metabasites** (Gil Ibarguchi & Julivert, 1988; Julivert & Durán, 1990a, b), but the tectono-metamorphic overprint on these mostly Lower Palaeozoic rocks is commonly too great to allow specific age determination.

The mostly siliciclastic Upper Ordovician succession of the CCR (Gutiérrez-Marco *et al.*, 2002) was deposited during northward movement away from polar regions to around 50°S by the end of Ordovician times, broadly equivalent today to a journey from Antarctica to Patagonia. This northward movement continued throughout Silurian times, when a pelagic sequence, comprising mostly graptolitic black shales and cherts overlain by limestones, records deep water conditions across a change towards warmer subtropical latitudes (*c.* 35°S) reached by Early Devonian times (Robardet & Gutiérrez-Marco, 2002; García-Alcalde *et al.*, 2002). Once again comparing this journey with modern geography, Silurian northward drift of what is now NE Iberia would represent movement from Patagonia to central Chile. Pelagic (mostly limestone) sedimentation continued into Devonian and earliest Carboniferous times (Table 1), with the area now closing in on Laurussia and drifting into low equatorial latitudes. The long period of stable, slow, pelagic sedimentation was finally terminated in Viséan times by the influx of immature, turbiditic sediments (mostly dark "**Culm facies**" sandstones) during the onset of the Variscan Orogeny (Colmenero *et al.*, 2002).

The Variscan Orogeny records the long period of diachronous oblique collision and strike-slip interaction between the Palaeozoic continents of Gondwana, Laurentia and Baltica and a number of continental microplates. In the Barcelona area it is marked by a break in sedimentation from Mid-Carboniferous to

Overview

Triassic times (a gap of around 80 **Ma**), and the deformation of the Palaeozoic succession. During this time, the world's continental plates assembled to form the Pangean Supercontinent, with what was to become Iberia lying around the equator within a continental **terrane** collage positioned between Gondwana and Eurasia (López-Gómez *et al.*, 2002). Late-orogenic intrusion of granitoid plutons was widespread in what are now the Pyrenees and the CCR (Castro *et al.*, 2002). In the CCR these intrusions form an enormous, composite, mostly **calc-alkaline** granodioritic batholith (Table 1) which crops out over around 1,500 km^2 from Barcelona (Montnegre Massif) and Montseny (Montseny-Guilleries Massif) to the northeast coast beyond Girona (Gavarres-Costa Brava Massif). This batholith was intruded in Early Permian times, with **radiometric ages** from the Montnegre Massif revealing subsequent cooling from 291 Ma to 276 Ma (Castro *et al.*, 2002; Solé *et al.*, 2002; López-Gómez *et al.*, 2002). It was intruded by a passive **stoping** mechanism, with steep walls and flat roofs, accompanied by widespread dyking, veining and contact metamorphism. In the Collserola area behind Barcelona metapelites, **metacalcareous** and metabasite mineralogies in the Palaeozoic country rock demonstrate a well-developed **metamorphic aureole** which reaches pyroxene **hornfels** facies adjacent to the granitoid intrusive contact, where corundum-spinel-sillimanite assemblages have been estimated to record **PT conditions** of around 1.5 kbar and 700°C (Gil Ibarguchi & Julivert, 1988).

Post-Variscan sedimentation in the CCR is represented in Barcelona by a small outlier of red alluvial sediments attributed to the Lower Triassic *Buntsandstein* facies (Table 1). Sediments belonging to this facies were deposited widely across Europe in continental extensional basins prior to the marine encroachment of the Tethys Ocean from the east in mid-Triassic times (López-Gómez *et al.*, 2002). To the north of the Barcelona area the east-west precursor to what was to become the Ebro Basin developed in response to the major post-orogenic palaeogeographic reorganisation which characterised Late Permian-Early Triassic times. Although preservation of Triassic rocks in Barcelona is limited to the small outlier previously mentioned, elsewhere in the CCR there are well-developed Upper Permian to Lower Triassic *Buntsandstein* successions resting unconformably on Palaeozoic basement, notable SW of the city (Prades-Priorat and Garraf areas), and behind the Littoral Range from the Llobregat Valley to Montseny.

As already stated, exposures of post-Triassic Mesozoic rocks are completely lacking in Barcelona, unlike along the coast to the SW, where Jurassic-Cretaceous carbonate sediments rest conformably above the Triassic succession (Fig. 2). These carbonate successions were deposited around a low-lying "Catalonian Massif" (now the area between Barcelona and the NE Catalan coast) as Iberia continued to drift through subtropical latitudes (25–35°N). The Iberian Plate had by now begun to act independently, rotating anticlockwise and moving SE, with a marine trough separating it from the European plate to the north and a wide marine carbonate platform and trough to the south connecting the Tethys Ocean with the opening Atlantic (Aurell *et al.*, 2002; Martín-Chivelet *et al.*, 2002). During Cretaceous times the Catalonian Massif formed a landmass defining the northwestern margin of the Tethys Ocean and was bordered by a major extensional, low-angle fault zone. This ancient tectonic boundary would later act as an important focus for Cenozoic movements induced by interactions between Africa and Europe (Table 1), repeatedly reactivating fault surfaces during phases of compression and extension.

Onset of Palaeogene northward-directed compression in NE Iberia resulted in Pyrenean uplift and, further south, the growth of an intraplate mountain belt (= ancestral CCR) during **transpressional inversion** of the Mesozoic extensional faults. Thrusting in the ancestral CCR began in the NE and moved SW over a period of at least 17 Ma, uplifting mountains along a zone 200 km in length as stresses were transmitted southwards from the Pyrenees. The rise of these ancestral mountains, which have been estimated to have reached a maximum height of nearly 4 km (Gaspar-Escribano *et al.*, 2004), resulted in the erosion under humid, sub-tropical climate conditions of much of the Mesozoic cover and part of the Palaeozoic basement. The resulting **synorogenic** erosional detritus was deposited as clastic sediment on either side of the range in the Ebro Basin and the offshore Barcelona Basin (López-Blanco *et al.*, 2000). Whereas the Barcelona Basin preserves an almost continuous sedimentary sequence from Late Eocene to present (lying above thick Mesozoic deposits), the sedimentary history of the Ebro Basin is more complex.

The contractional boundary between the Ebro Basin and the CCR is well exposed around the Llobregat Valley, adjacent to the Vallès-Penedès Fault some 30 km NW of Barcelona, where NW-verging folds and thrusts deform both Palaeozoic basement and Mesozoic cover. The added complication of accentuated local uplift due to transverse NW-SE faulting in this area produced a highly active sedimentary point source from which were shed thick Eocene conglomeratic sequences now spectacularly exposed on Montserrat (Roca *et al.*, 1999; López Blanco *et al.*, 2000). Within Barcelona city, although no Palaeogene rocks are present, thrust-**imbrication** of the Palaeozoic basement can in places be demonstrated to involve post-Variscan (Triassic) sediments and is therefore interpreted as Palaeogene in age. By the end of this compressional phase, which terminated in mid-Oligocene times, it has been calculated that around 1.3 km of rock had been removed by erosion, and the CCR had subsided **isostatically** by another 1.2 km, leaving a mountain chain 1,200–1,900 m high in the Barcelona area (Gaspar-Escribano *et al.*, 2004).

During the Oligocene epoch the Eurasian and Iberian plates became welded (Table 1) and tectonic activity refocussed further south into the developing Betic-Balearic mountain belt. In NE Iberia, the compressional regime that had controlled the tectonic setting underwent an abrupt reversal during Late Oligocene onset of rifting and opening of the Valencia Trough. Initially rapid extension along SE-dipping listric faults over a period of around 5 Ma (Late Oligocene-Early Miocene) produced the **horst-and-graben** pattern that dominates the present-day structure of the CCR (Fig. 2 cross section; Roca & Guimerà, 1992; Roca *et al*., 1999). Extensional fault subsidence continued into Late Miocene times, although at a slower rate, and the countering effects of **flexural uplift** aided further erosion of the CCR by river systems shedding sediments into coastal deltas. The total amount of extension in the CCR over the Late Oligocene-Miocene period has been estimated at 15.5 km (10.8 km on the Barcelona Fault and 4.7 km on the Vallès-Penedès Fault: Gaspar-Escribano *et al*., 2004). During this extensional collapse of the CCR, continental to shallow-marine Miocene sediments were deposited within the Vallès-Penedès **half-graben** lying between the Prelittoral and Littoral ranges. In front of the Littoral Range, parallel to the modern Mediterranean coastline, a thick succession of mostly marine Miocene sediments was deposited on the hanging wall of the Barcelona Fault. In the shallow coastal waters close to land a siliciclastic deltaic sequence was laid down, and part of this sedimentary succession (Tables 1 and 2) is now exposed on Montjuïc Hill in Barcelona. These sediments are capped by a regional erosion surface produced during the Messinian salinity crisis when contact with the Atlantic Ocean was temporarily lost and most of the Mediterranean Sea evaporated.

Post-Messinian marine transgression (Table 2) resulted in the deposition of argillaceous marine Pliocene sediments around the flanks of the CCR. These Pliocene blue clays are present beneath much of the lower parts of Barcelona city and they grade up into a more sandy regressive sequence overlain in turn by a varied succession of Quaternary sediments. The Pleistocene succession covering the modern coastal plain includes red clays, nodular **palaeosols**, **loess** and gravelly wash (Table 2). Coastal sediments deposited offshore from Catalonia at this time record the vertical stacking of seaward **downlapping** regressive deposits, a process controlled by **eustatic** sea-level fall and seaward coastal progradation (Ercilla *et al*., 1994). Above these materials flowed the torrential ephemeral streams known as *riadas*, depositing sands, gravels, and channelised coarser deposits full of debris from the hills behind the city. Further seaward, Holocene deltaic sediments spread southwest from the Besòs river estuary over the last 10,000 years and are now overlain by the modern city beach. On the other side of the city the much larger delta deposited by the Llobregat river shows Pliocene blue clays overlain by Pleistocene fluvial gravels (up to around 20,000 years old), above which are up to 64 m of Holocene delta sands, silts and clays (Gámez *et al*., 2005; Lafuerza *et al*., 2005; Simó *et al*., 2005; Vázquez- Suñé *et al*., 2006).

Holocene sediment accumulation on the Llobregat delta was initially slow (1 mm/yr) during rapid Holocene transgression and low clastic supply, but increased (12–25 mm/yr) after around 5,000 years ago (Gámez *et al*., 2005). At this time the climate of the Barcelona area improved from cooler conditions towards the current tendency for hot, dry summers, encouraging a change in vegetation cover with evergreens and shrubs replacing deciduous forest, and thus a change in human agricultural habits towards exploiting summer mountain pastures by transhumance (Jalut *et al*., 2000) (Table 2). The current three-month dry

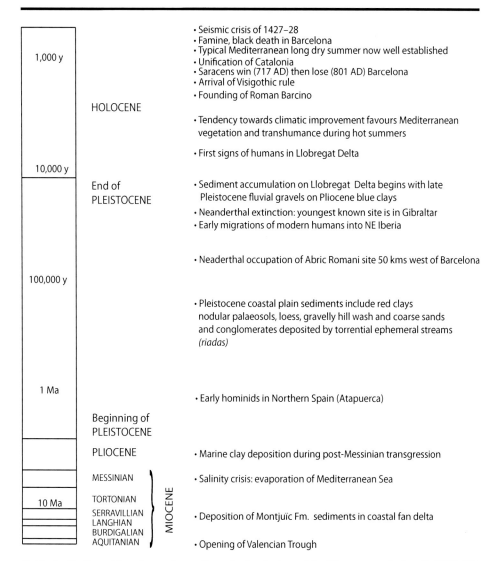

1,000 y		• Seismic crisis of 1427–28 • Famine, black death in Barcelona • Typical Mediterranean long dry summer now well established • Unification of Catalonia • Saracens win (717 AD) then lose (801 AD) Barcelona • Arrival of Visigothic rule • Founding of Roman Barcino
	HOLOCENE	• Tendency towards climatic improvement favours Mediterranean vegetation and transhumance during hot summers • First signs of humans in Llobregat Delta
10,000 y	End of PLEISTOCENE	• Sediment accumulation on Llobregat Delta begins with late Pleistocene fluvial gravels on Pliocene blue clays • Neanderthal extinction: youngest known site is in Gibraltar • Early migrations of modern humans into NE Iberia • Neaderthal occupation of Abric Romani site 50 kms west of Barcelona
100,000 y		• Pleistocene coastal plain sediments include red clays nodular palaeosols, loess, gravelly hill wash and coarse sands and conglomerates deposited by torrential ephemeral streams *(riadas)*
1 Ma	Beginning of PLEISTOCENE	• Early hominids in Northern Spain (Atapuerca)
	PLIOCENE	• Marine clay deposition during post-Messinian transgression
	MESSINIAN	• Salinity crisis: evaporation of Mediterranean Sea
10 Ma	TORTONIAN SERRAVILLIAN LANGHIAN BURDIGALIAN AQUITANIAN } MIOCENE	• Deposition of Montjuïc Fm. sediments in coastal fan delta • Opening of Valencian Trough

Table 2. Stratigraphic column drawn from the beginning of the Neogene period until 1500 AD, listing the sedimentary rocks on which most of Barcelona has been built. Relevant geological, archaeological and historical events are included. The Neogene Period is subdivided into the Miocene and Pliocene epochs. Also shown are the six ages of the Miocene epoch, during one of which (Serravillian) the Montjuïc Formation sandstones were deposited. Note the use of logarithmic scale.

13

summer typical of Barcelona has been the norm only for the last 1,000 years or so, with both pollen data and historical sources recording an increase in olive cultivation (Jalut *et al.*, 2000). Two prominent recent peaks of sediment accumulation on the Llobregat delta have been attributed at least in part to pulses of increased anthropogenic deforestation, one in the Visigothic period (6th–8th century) to the Upper Medieval Age (10th–14th century), the other in the 18th century during economic recuperation after the depopulation crisis of the Lower Middle Ages (1350–1487 AD; Gámez *et al.*, 2005).

EXCURSIONS

The three excursions described below move from the oldest rocks exposed behind Barcelona then down through the city eventually to the youngest rocks underlying the coastal zone. The excursions are interlinked so that the first one finishes where the second one begins and the third excursion begins after a short metro ride from where the second one ends (Fig. 4). Although it is possible to complete the first two excursions in one exceptionally long day, a more relaxed approach would treat each one as a separate trip. The last excursion includes much of historical and cultural interest (Roman ruins, Medieval churches, museums and parks) and deserves a full day. No special equipment is required, although a compass and hand lens are always useful. As much of the route involves city walking, carrying a hammer is not appropriate, and heavy field boots are not necessary. All excursions are based on using cheap public transport and a private vehicle will be far more trouble than it is worth in this busy, densely populated and traffic-polluted city. The metro, FGC train, funicular and bus systems are integrated within the Transports Metropolitans de Barcelona (TMB) system, so that one multiperson T10 ticket (in 2011 still <10 euros) allows 10 journeys on all routes. All the usual warnings about petty thieving in European cities apply, not least because the route passes through some of the most tourist-visited parts of the city (Park Güell, Las Ramblas, Montjuïc and metro line 3). Even geologists are not immune to pickpockets, camera theft and bag snatching.

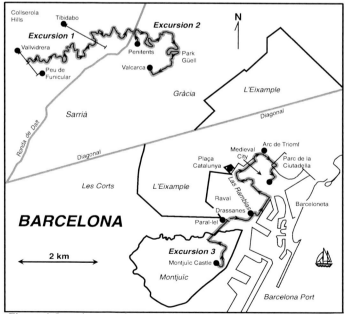

Figure 4. Location map for Excursions 1–3.

Excursion 1: Passeig de les Aigües

EXCURSION 1: PASSEIG DE LES AIGÜES

This first excursion makes use of a track known as the Passeig de les Aigües which contours around the Collserola Hills high above Barcelona and is easily reached via a funicular integrated within the TMB public transport system. Apart from offering spectacular views of the city and its surroundings, the Passeig also allows examination of the metamorphic and igneous Palaeozoic basement rocks that typify much of the Catalan Coastal Ranges (CCR). The 7–8 km walk is easy and can be completed in 3–4 hours, including stops. Occasional deviations from the generally flat track involve a minor amount of scrambling up and down gravelly hill paths, so footwear with a good grip and ankle protection is recommended. The route is relatively little known to tourists, although it does get busy with cyclists, joggers and walkers, especially at weekends and on weekday evenings. Despite the proximity of the track to urban areas, wild boar can be encountered (usually in the early morning and evening) but are highly unlikely to present any problem. Much more serious is the need for sun protection during much of the year and there are no services en route so food and water need to be carried. The route is best avoided if thunderstorms are forecast because once started there are few escape routes.

To begin the excursion, use a T10 ticket (Zone 1) to take the Sabadell (S2) or Terrassa (S1) FGC train from Plaça de Catalunya to Peu de Funicular (5 stops, avoid using the last train carriage as the platform is very short). Take the lift (or steps) from the platform up to the Vallvidrera funicular. Once in the funicular car, press the button for Carretera (Passeig) de les Aigües, which programmes a stop halfway up the ascent. The exit from the funicular at this stop leads to a wide track (the Carretera or Passeig de les Aigües) where we turn left (Fig. 5).

Vallvidrera funicular to Serra de Vilana
Walk for 5 minutes along the level track which curves through pine trees, merges with lanes from the left and right, becomes a paved road, and passes a garage on the left. Just beyond the garage are the first good exposures of the metamorphic sequence which underlies most of the Collserola Hills (Location 1 on Fig. 5). Strongly foliated, fine-grained, locally quartz-veined metabasites and metasediments show an S_1 **foliation** dipping north at around 30–50°. The **protoliths** to these unfossiliferous rocks are, by analogy with other basement massifs in northern Iberia, interpreted as Lower Palaeozoic in age (see Stratigraphic Overview section above and Gutiérrez-Marco *et al.*, 2002).

Figure 5. (Facing page) Route map for Excursion 1: Passeig de les Aigües. The route begins on low grade regional metamorphic basement and moves eastwards into increasingly contact metamorphosed rocks in a dyke/vein complex adjacent to a granodioritic pluton.

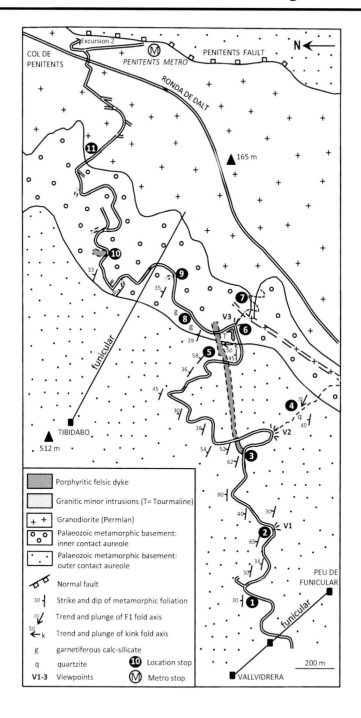

COL DE PENITENTS

Excursion 2

PENITENTS METRO

PENITENTS FAULT

RONDA DE DALT

N

▲ 165 m

11

10

33

80

35

9

8

g

g

39

7

V3

T

6

58

5

50

45

36

45

4

30

q

40

20

38

V2

TIBIDABO

54

52

3

62

▲ 512 m

funicular

80

40

30

2

V1

65

34

30

PEU DE FUNICULAR

30

1

funicular

VALLVIDRERA

200 m

Legend:

- Porphyritic felsic dyke
- Granitic minor intrusions (T= Tourmaline)
- Granodiorite (Permian)
- Palaeozoic metamorphic basement: inner contact aureole
- Palaeozoic metamorphic basement: outer contact aureole
- Normal fault
- 30 ⌐ Strike and dip of metamorphic foliation
- 20 ↙ Trend and plunge of F1 fold axis
- 50 ←k Trend and plunge of kink fold axis
- g garnetiferous calc-silicate
- q quartzite
- **10** Location stop
- V1-3 Viewpoints
- Ⓜ Metro stop

17

Excursion 1: Passeig de les Aigües

Continue following the Passeig de les Aigües which curves right at Plaça de la Font del Mont (where a lane climbs up to the left), around the headwaters of one of the many ephemeral streams (*rieras*) that cascade down into the city during times of heavy rainfall. There are exposures of the same north-dipping metamorphic sequence on the left, although locally the dip steepens to around 70° due to flexure over **monoclinal folds** with E-W trending axes. The track curves left then right, with extensive views opening out on the right, and passes more of the same metamorphic sequence before curving left around a prominent viewpoint (Fig. 6) marked by a wall on the right (Location 2/Viewpoint 1 on Fig. 5).

The view here overlooks much of the modern city, most of which is built on Pleistocene sediments that slope down to the coastal zone where recent debris eroded from the Collserola Hills coalesces with deltaic materials washed into the sea from the Besòs river (out of sight to the left: see Fig. 2). The twin towers of Port Olímpic (left) rise from this coastal zone behind the city centre. Ahead, the view overlooks the Miocene sandstone hill of Montjuïc (Fig. 6), which rises to 192 m and is described in Excursion 3. Further to the west is the low-lying Llobregat delta, a site of human activity since at least Neolithic times, with extensive urbanisation these days spreading out towards El Prat airport at the coast (El Prat = Catalan for meadow). In Roman times the delta shoreline lay up to 2 km further inland than today, allowing boat shelter on the western side of Montjuïc (Vázquez-Suñé et al., 2006). Further to the right of the modern delta rises the peak of Sant Pere Martir (389 m), capped by a telecommunications tower at the SW end of the Collserola Hills. Sant Pere Martir is underlain by a Permian granodiorite which has intruded the Palaeozoic metamorphic sequence by passive stoping. The NE continuation of this same **pluton** provides the main theme of the geology seen during this excursion, which visits granitic dykes emanating from the pluton, a zone of granite veining and intense metamorphism close to the contact and finally the outcrop of the granodiorite itself.

From Viewpoint 1 the Passeig de les Aigües turns NE, passing further exposures of metamorphic basement on the left to enter more open shrubland beneath the huge telecommunications Torre de Collserola, and contours around another *riera* valley draining south towards the Llobregat delta. Around 500 m from Location 2 there is a metal font just before a right bend (Location 3, Fig. 5). The exposures on the left here, by a red and white post just before the metal font, reveal a weathered, pale, **porphyritic** dyke 6–9 m thick, with **phenocrystic** quartz, feldspar and biotite lying in a fine **felsic** matrix. These dykes are interpreted as magmatic offshoots from the granodioritic pluton, the contact with which lies <1 km to the south, and their distinctive porphyritic texture has proved useful in identifying the provenance of detrital clasts in the Miocene sediments of Montjuïc (Excursion 3). On the other side of the font are more exposures of the metamorphic country rock. These show thin, grey calcareous layers within the more typically pelitic sequence. The S_1 foliation here is folded by minor **kinks** with N-plunging axes.

Figure 6. View southeast over Barcelona and Montjuïc from Location 2, Excursion 1, from the twin towers of Port Olímpic (left) to the eastern margin of the Llobregat Delta (right). The southern end of the city, constricted between Montjuïc and the Collserola Hills, is built mostly on Plio-Pleistocene sediments.

Continuing another 150 m from Location 3 the track curves left around another prominent outlook (Viewpoint 2), turning a hairpin bend around the projecting ridge of the Serra de Vilana. Here the views are even more expansive, sweeping clockwise from the Torre de Collserola to the Fabra Astronomical Observatory and down across the city. At this point leave the main track and follow the small path that continues SE along the projecting ridge of the Serra de Vilana, initially down then climbing gently to reach (after around 150 m) exposures of quartzite. A little further on, passing the hilltop on the left, there is a rare exposure (left) of an F_1 upright **antiform** in the quartzite, with a strong steep S_1 **axial planar fabric** defined by a spaced foliation and thin, white quartz veins (Fig. 7; Location 4 on Fig. 5). The fold hinge and associated L_1 intersection **lineation** plunge gently to the NW, in the direction of the hilltop suburb of Vallvidrera.

Excursion 1: Passeig de les Aigües

Figure 7. Gently NW-plunging upright antiformal F_1 fold hinge in Palaeozoic metaquartzite of the Serra de Vilana, looking NW (Excursion 1, Location 4). Note well-developed L_1 intersection lineation (S_0/S_1) on bedding planes in the fold core (arrow), and thin white quartz veins lying parallel to sub-vertical S_1.

Serra de Vilana to Tibidabo funicular

Return to the main track which continues around the east flank of the Serra de Vilana, passing the continuation of the pale porphyritic dyke seen at Location 3 (weathered and easily missed exposures across an area around 10 m wide). Beyond this the track continues to a right curve, passing exposures of metasediments in some of which one can discern the slaty fabric overprinted by contact metamorphic **porphyroblastic** "spots" (best noted in loose fragments by the track side, although much better examples will be seen later). The track continues winding around the hillside, passing a bike/running circuit signpost (2 km) and numerous further exposures of the north-dipping, intensely foliated basement rocks around a series of bends. Pass another metal font at a sharp right bend in another *riera* valley head, and continue past signpost 1.5 km to pass a wide track leading down to the right (signposted CO3), just beyond which on the left there is another exposure of the porphyritic dyke (Location 5 on Fig. 5). The **discordant** intrusive contact of this undeformed dyke intruding kinked metasediments is well seen here.

Figure 8. Tourmaline-bearing (arrow) late stage granite vein intruding contact-altered Palaeozoic metapelites. Excursion 1, Location 6.

Continue on to where the track curves left then passes signpost 1 km after which there is another tremendous view over the city (Viewpoint 3/Location 6 on Fig. 5). On the left side of the track here a white, tourmaline-bearing, composite layered **aplitic/pegmatitic** granite vein <1 m wide intrudes contact-altered meta-sediments (Fig. 8). The textural changes in grain size and mineral layering ex-hibited by this vein reflect complex crystallisation patterns during rapid cooling. Such veins are likely to cool in days or months rather than years, and the large size of many crystals may be due to rapid growth promoted by localised concen-trations of incompatible "flux" elements like boron and fluorine (see discussion of this interesting topic in London, 2009). This late magmatic stage minor intru-sion, which locally cross cuts the S_1 metamorphic foliation, is one of several in this area, forming part of a vein-dyke complex in the inner part of the plutonic contact aureole. The growth of cordierite porphyroblasts in response to the con-tact metamorphism of more pelitic layers is becoming more clearly evident (see metapelite in lower part of Fig. 8).

There are particularly good views from here northeast over the group of wooded hills which rise from the urban landscape and are known as Les Turons (Figs 9 and 10). These hills are separated from the Collserola Hills by a shallow

Excursion 1: Passeig de les Aigües

Figure 9. View northeast from Excursion 1, Location 6, across the Barcelona hills known as Les Turons (visited in Excursion 2). In the centre is the quarried hilltop of La Creueta del Coll (limestone), to the right of which the Turó del Carmel rises to a height of 267m asl. Between these two hills is the valley suburb of El Coll, whereas to the far right are the wooded slopes behind Park Güell. To the left of La Creueta del Coll is the valley watershed of the Col de Penitents, through which burrows the urban motorway of the Ronda de Dalt (seen lower right). The turreted mansion to the left of the Col de Penitents is built on the Collserola slopes immediately above the Tibidabo funicular station. Behind in the distance spreads the densely urbanised coastal strip running NE towards the Costa Brava, where the granodioritic hills of the Montnegre Massif (seen top left) meet the sea (Fig. 2). The photo was taken during a winter anticyclonic air pollution episode and the resulting grey contaminated atmospheric mixing layer can be seen hanging over the city.

valley (Col de Penitents) through which burrows the Ronda de Dalt urban motorway and where Excursion 1 finishes and Excursion 2 begins (Fig. 5). There are four main wooded hills as viewed from here: that to the left is La Creueta del Coll (with a prominent quarry face), further to the right rises the highest point on Turó del Carmel (267 m), to the right of which are the thickly wooded slopes behind Park Güell (Fig. 9). Further to the right the smaller hill of Putget is isolated from Park Güell by the densely urbanised valley of the Vallcarca *riera* (Fig. 10). Each of these hills comprises massifs of Palaeozoic sediments within a fault complex which borders the SE side of the Collserola Hills (Figs 2 and 5). The view over the city centre to the right of Putget overlooks the 19th–early 20th century part of

Excursion 1: Passeig de les Aigües

Figure 10. View looking east from Excursion 1, Location 6. The wooded slopes behind Park Güell (already noted in the previous figure) rise on the left. The smaller elongate hill on the right is Putget, a fault-bounded "pip" of Palaeozoic basement within Les Turons Fault Zone. Between the two hills runs the densely urbanised Vallcarca *riera* valley, opening out eastwards towards the city centre in the direction of Gaudí's La Sagrada Familia. The city beyond Les Turons is built upon the low, gently sloping Pleistocene platform known as the *Pla Alt*. Also seen is the distinctive form of the Agbar Tower from which recent high-rise development runs left down to the sea, tracing the route of the Diagonal highway which slices through the entire city (see map of excursion routes). To the right are the twin towers of Port Olímpic, seen previously in Figure 6.

the city known as L'Eixample ("the enlargement"), with little green space alleviating the densely grid-packed urbanisation. Returning to the view over the Col de Penitents, on a clear day from here can be seen the granodioritic hills of the Montnegre Massif (Littoral Range) running northeast out to the Costa Brava and confining urban development within the narrow coastal strip.

It is worth exploring the ground below Viewpoint 3 by taking the narrow path signposted to Bonanova. This path descends across instructive exposures of the intrusive complex of veins, dykes and sills which characterises the inner envelope of the granodioritic pluton. A short distance down the path is a granodioritic dyke striking NE and sharply discordant to the foliated country rock metasediments. The dyke margins are themselves intruded by the white tourmaline granite veins seen above at the viewpoint. Descending further, passing a small cactus

23

Excursion 1: Passeig de les Aigües

Figure 11. Irregularly veined upper margin of biotite granodiorite intruding contact-altered Palaeozoic metapelites. Excursion 1, Location 7. Pen is 15 cm long.

field, the path turns sharply right after two old pillars, with views down (left) over the Ronda de Dalt motorway. A few metres further down the path crosses another minor granodioritic intrusion, this time intruding the country rock as a sill-like body, parallel to the metamorphic foliation. Continue down the hillside following the path, which turns left and passes through more cactus to a sharp right bend where the irregularly veined upper margin of the sill (Fig. 11) is well exposed in the path (Location 7 on Fig. 5). A few metres below here along the path the weathered granodiorite is unusually rich in **euhedral** biotites displaying perfect hexagonal basal sections. From here at Location 7 climb back up to viewpoint 3.

The Passeig de les Aigües now continues northwards, passing more examples of granitic veins, back across the poorly exposed granitic dyke seen at Location 5 (Fig. 5). After passing a small path descending to the right and then a paved shallow drainage course crossing the Passeig, we enter an area where there are good exposures on the left of north-dipping massive and banded calc-silicates displaying pale pink garnetites containing dark clinopyroxene-bearing lenses and layers (Location 8, Fig. 12). The garnets, some of which reach centimetres in size, are CaFe-rich (**grossular-andradites**), and the pyroxenes show a mixed composition dominated by diopside-hedenburgite (Gil Ibarguchi & Julivert, 1988). The calcareous and calc-silicate lithologies present within the con-

Figure 12. Pale garnetiferous and dark clinopyroxene-bearing (CPX) layers in contact-altered Palaeozoic calc-silicates. Excursion 1, Location 8. Coin is 2.4 cm diameter.

tact aureole in Collserola exhibit a distinctive and complex mineralogy which includes chlorite, biotite, Ca-amphibole, epidote, Ca-pyroxene, idocrase, garnet and scapolite. This contrasts with the contact-metamorphic porphyroblasts in the metapelites which are mostly biotite, cordierite and (in black shale protoliths) andalusite. Only in the highest-grade areas (a small zone in this area just SW of the Tibidabo Funicular, as mapped by Gil Ibarguchi & Julivert, 1988) has the assemblage sillimanite, corundum and scapolite been recorded in metasediments, along with Ca-pyroxene in metabasites.

The track continues curving left then right around another minor valley, passing another font. The highest-grade metamorphic conditions in these innermost parts of the granodioritic contact aureole have been estimated to have reached 700°C and 1.5 kbar (Gil Ibarguchi & Julivert, 1988). At signpost 0.5 km, just over halfway between the font and the next (left) bend, there is an exposure of small, 2-mica-tourmaline granitic dykes intruding coarsely spotted pelitic hornfels (Location 9, Fig. 13). The sporadic exposures on the left beyond here are metapelites also coarsely spotted by contact-metamorphic minerals (notably cordierite). Beyond this Location, the Passeig de les Aigües curves left, passing more coarse hornfelses and a track (left) descending from Tibidabo, to cross the Tibidabo funicular.

Excursion 1: Passeig de les Aigües

Figure 13. Tourmaline + 2-mica granite dyke intruding coarsely spotted (arrows) cordierite metapelite in the inner aureole of the granodiorite. Excursion 1, Location 9. Coin is 2.3 cm diameter.

Tibidabo funicular to Col de Penitents

From the funicular the track follows a series of bends, passing a path on the left (along which there are further exposures of contact-altered country rock intruded by a granitic dyke) to a car park where a road leads down to Tibidabo funicular station. Stay left of the car park, contouring horizontally past signpost 0 km on the Passeig de les Aigües, which becomes paved and curves past further exposures of north-dipping spotted metasediments on the left, to reach a long left bend finishing at a concrete pillar where there is a particularly good exposure of a porphyritic granitic dyke (Location 10).

Beyond this again one enters the dyke-vein complex invading highly recrystallised contact-altered country rock, well exposed on the left just where a road (Carrer del Maduixer) descends towards the city. Leave the Passeig de les Aigües here and take the Carrer del Maduixer, descending the urbanised hillside to pass more exposures of metasedimentary country rock intruded by granitic dykes (seen opposite No. 34–36 at a sharp right turn). After this the road becomes Carrer de Vallpar and descends further to reach a bus stop (route 124). Here a road descends to the right (but don't take it) and there are further exposures of dyke-intruded country rock on the left. The irregular and commonly faulted granitic dykes can be seen enveloping **xenoliths** of baked country rock in a style typical of the passive stoping intrusion mechanism which characterises the Variscan plutons of the CCR. Ignoring the road to the right, continue ahead (SE) down the narrow lane. Exposures on the left mark the change from metasediment to granodiorite as the pluton is reached (Location 11: just before the first house on the right) and the rest of the route lies exclusively on the granodiorite outcrop.

Continue down Carrer de Vallpar, passing further exposures of granodiorite on the left, to another bus stop and a fork in the road. Here, keep left to pass another small road on the right and follow a long left bend past Navata 26–34 to a straight descent, passing roads to the left (Carrer de Trullols) and right (Carrer de Sant Onofre) to traffic lights at a T-junction with a main road (Carretera de Sant Cugat). Cross the main road to locate several flights of steps on the right which descend to another set of traffic lights in the Col de Penitents watershed area lying between the Collserola Hills and Les Turons. Cross the lights, passing over the Ronda de Dalt motorway. This is the end of the excursion. From here either join Excursion 2 or head home on the Metro. To reach the Metro (Penitents station) continue ahead down steps to turn right into Avinguda de Vallcarca and walk downhill for 5 minutes. Alternatively, to join Excursion 2, turn left across the zebra crossing to locate Carrer de Veciana on the Col de Penitents (Figs 5 and 14, top left).

Excursion 2: Les Turons

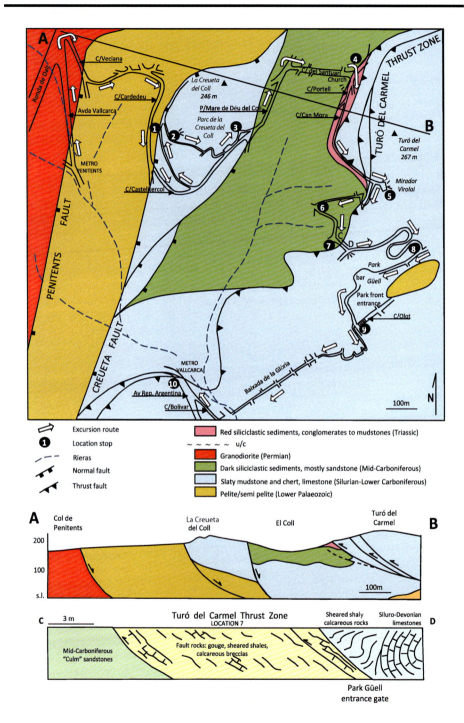

EXCURSION 2: LES TURONS

This excursion describes a 6 km (2–4 hour) walk from Penitents in the north-west of the city to Park Güell, one of the most-frequented tourist attractions in Barcelona (and therefore best avoided at weekends). The walk focuses on the Palaeozoic and Mesozoic geology of the hills known as Les Turons (note Les in Catalan is pronounced Las), starting on the oldest rocks in the city (Cambro-Ordovician metasediments) and moving up through representatives of the overlying Siluro-Devonian pelagic sedimentary sequence and the later deposits of coarser, Carboniferous turbiditic sediments which are the youngest Palaeozoic sediments exposed in the region. The route provides extensive views across Barcelona and also allows a visit to the only area in the city limits where Mesozoic rocks are exposed. The major theme, however, is more structural than stratigraphic, as all rocks have been so extensively disrupted by normal faults and thrusts that the entire outcrop of Les Turons is viewed as lying within a fault complex which records both Palaeogene compression and Neogene extension. The engineering challenges arising from excavating underground through such faulted ground were dramatically illustrated in 2005 by a major collapse of residential flats into a metro line being tunnelled through the hills.

Penitents to La Creueta del Coll

This first leg involves an undemanding 1.5 km urban walk through the north-western corner of the District of Gràcia to La Creueta del Coll, a former quarry converted into a public park and the first of Les Turons to be visited during the excursion. Geologically it continues on from the previous excursion along Passeig de les Aigües, starting on the southeastern margin of the same Collserola basement block visited earlier (Figs 5 and 14).

Ride the (green) metro line 3 to Penitents, 4 km NW of Plaça de Catalunya. Leave the metro station by escalators, take the left exit (Av. de Vallcarca) and walk uphill northward along busy Avinguda de Vallcarca (Fig. 14) which climbs the valley head of the *Riera* de Vallcarca, an ephemeral stream draining the Collserola slopes immediately to the north. Prior to urbanisation this *riera* fed into one of the principal drainage systems of the city, running southeast at the surface to the Medieval city (Sanz Parera, 1988; Riba & Colombo, 2009).

Figure 14. (Facing page) Simplified geological map and cross sections for Excursion 2. The route moves from the granodiorite outcrop at Col de Penitents (A in cross section: where Excursion 1 ends) across a prominent normal fault in the hanging wall of which are Lower Palaeozoic low grade metasediments (see Fig. 15). Further east are faulted slices of Silurian-Lower Carboniferous low grade metasediments, locally overlain by Triassic red beds. Stratigraphic coherence within the Palaeozoic sequence has been lost by poor exposure and pervasive faulting, both extensional and compressional (thrusts). Much, and probably all, of the faulting is attributed to Alpine (Cenozoic) deformation. Cross section A-B broadly follows the first half of the described route. Cross section C-D (Location 7) illustrates the thrust zone exposed in a road cutting at the NW entrance to Park Güell.

Excursion 2: Les Turons

After 400 m there are traffic lights adjacent to the Ronda de Dalt, an urban motorway that cuts NE-SW behind the city between the Llobregat and Besòs rivers, at the foot of the Collserola Hills (Figs 5 and 14). Here we have climbed to around 150 m above sea level to reach the Col de Penitents, a watershed separating the Vallcarca and Horta *riera* drainage systems, and there are wide views west to the Collserola peaks of El Tibidabo (512 m) and northwest to Turó de Santa Maria (446 m). The hills here represent a continuation of the geology seen in Excursion 1: Palaeozoic metasediments intruded by Permian biotite granodiorite (Fig. 5). This point is on the eastern limit of the granodiorite outcrop, close to where the Penitents Fault has brought down Cambro-Ordovician metasediments against the intrusion (Fig. 14). This tectonic boundary forms part of a major, east-dipping Neogene normal fault system running parallel to the Collserola Hills, curving broadly NE-SW under Pleistocene-Holocene sediments across the entire city from the river Besòs in the NE to Espluges de Llobregat in the SW (IGC/ICC, 2009a).

Turn sharp right after the lights into Carrer de Veciana which initially climbs SSE and passes artificially covered cliff exposures below buildings (on the left). This is the eastern limit of the granodiorite outcrop. Immediately beyond this, just before forking to the left, we cross the unexposed Penitents Fault (Fig. 14) and there are small exposures of weathered Lower Palaeozoic low grade metasediments in the hanging wall high above the left side of the road. Curve left, still following Carrer de Veciana, noting the use of grey, medium-grained granodiorite and contact-metamorphosed grey metasediments as building stone in the walls on the left. After a further 200 m along Carrer de Veciana, having passed two roads to the left (allowing views northwest towards Turó de Santa Maria) we reach a complex road junction with the prominent hill of La Creueta del Coll rising ahead. Here bear right down Carrer de Cardedeu, which curves southwards with parkland to the left, and continue for around 250 m to house No. 51 (Location 1 on Fig. 14). Across the road from here are exposures of weathered, slaty, chlorite-grade Cambro-Ordovician pelitic and semipelitic metasediments showing the main S$_1$ foliation curving around a **recumbent fold** (Fig. 15).

Behind these deeply weathered exposures of Cambro-Ordovician rocks, the parkland rises steeply towards the summit of La Creueta del Coll. This hill exposes a Siluro-Devonian pelagic sedimentary sequence in the hanging wall of another east-dipping normal fault which is referred to in this guide as the Creueta Fault (Fig. 14). Continue south along Carrer de Cardedeu to take the left fork into Carrer de Castellterçol which curves left (eastward), crossing on to the hanging wall of the unexposed Creueta Fault (Fig. 14) to reach a narrow lane (Passatge de Manlleu) on the left after 175 m. Turn left into this lane to enter the undeveloped scrubby parkland behind the quarry of La Creueta del Coll.

Excursion 2: Les Turons

Figure 15. Recumbent fold (arrow) in low grade Palaeozoic sediments lying in the footwall of the Creueta Fault. Excursion 2, Location 1 (see Fig. 14).

Parc de la Creueta del Coll

There is abundant exposure of Palaeozoic limestone in the extensively quarried hill of La Creueta del Coll, and the overall dip is eastward so that the oldest strata are found here on the west side of the park. The first rocks encountered on the right along Passatge de Manlleu are east-dipping limestones correlated within the Siluro-Devonian La Creu Formation, the lowest part of which is Ludlow in age (Julivert & Durán, 1990a, b; Robardet & Gutiérrez-Marco, 2002; Plusquellec *et al.*, 2006). As is true elsewhere in Les Turons, however, extracting stratigraphic detail here is hampered by pervasive faulting. The faulted calcareous rocks in these exposures, low on the west side of the hill, are mostly highly fractured, crudely foliated and cut by pink **breccia** veins.

Climbing higher along the Passatge de Manlleu, ignoring a path to the right, views open out to the southwest over Collserola and the Llobregat delta. Here, along the western side of the hill, the geology comprises Silurian pelites which rest on the hanging wall of the Creueta Fault and lie stratigraphically below La Creu limestones. The existence of fine-grained, graptolitic Silurian pelagic sediments in the Barcelona area has been known for over 100 years (Barrois, 1893, 1901; Julivert *et al.*, 1985) and here they provide a suitably weak horizon lying directly on the hanging wall of the underlying normal fault (IGC/ICC,

Excursion 2: Les Turons

2009b). Given the soft and commonly tectonised nature of this pelitic unit and its transition into the overlying limestones, exposures are poor. Continuing on the main track (ignoring a path leading down to the left) a path comes in acutely from the right, adjacent to a low exposure of folded, foliated and weathered Silurian siliciclastic sediments (better exposures will be visited later). Climb this path on the right, after crossing a metal chain, and head south for a few hundred metres, ignoring a track on the left. Extensive views soon open out south over the city towards Turó del Putget, although for a view over the quarry and hills to the east, take the small path on the left which climbs briefly to a metal fence.

The view from the metal fence (Location 2) looks out northeast across the recreational park below to the limestone quarry face, and east across the suburb of El Coll to the peak of Turó del Carmel (267 m). Turó del Carmel exposes the same Siluro-Devonian sequence seen here in La Creueta del Coll, whereas much of the low ground of El Coll below is underlain by Mid-Carboniferous siliciclastic sediments. The contact between these two stratigraphic units is part of an imbricated thrust belt which is referred to as the Turó del Carmel Thrust Zone and which brings older Siluro-Devonian rocks over Mid-Carboniferous and Triassic sediments (Fig. 14). The fact that Triassic rocks are involved in the thrusting leads to the interpretation of this structure as being Alpine (Palaeogene compression) rather than Variscan. The lower part of the Turó del Carmel Thrust Zone runs just below the level of the prominent Escola Virolai and is expressed geomorphologically in the changing slope of the hillside. In contrast, the boundary between the turbiditic sediments of El Coll and the calcareous rocks of the Creueta del Coll quarry below is another prominent east-dipping normal fault, which is interpreted as a splay from the Creueta Fault (Fig. 14). Finally, to the right (south) of Turó del Carmel are the pine-forested slopes that lie immediately behind Park Güell, visited later during this excursion. These pine forests lie mostly on Carboniferous rocks, with the overlying thrust zone running along the higher part of the ridge. Behind these forested slopes rise the towers of the Sagrada Familia and Port Olímpic.

Descending a few steps from the metal fence lookout, immediately on the right there are poor exposures of weathered pale limestones intruded by brown carbonate veins. However, there are better exposures of these rocks further during the descent into the park below. Return to the main track and continue south (left), with views towards Montjuïc ahead in the distance and the smaller hill of Turó del Putget. Like the other two Turons previously mentioned, Putget is also preserved as parkland. It represents a continuation of the same geology, with limestones forming the highest point underlain by thrusts, although here the imbricated sequence lies directly on the hanging wall of the underlying normal fault system. Further to the right the southwesterly continuation of the Penitents Fault passes beneath the lower ground below Collserola, crossing the Barcelona Districts of Sarrià-Sant Gervasi and Les Corts and on out towards the Llobregat delta. Closer to hand, on the left side of the path, are low exposures of pale

Figure 16. View looking north in the Parc de la Creueta del Coll which lies within a former quarry exposing east-dipping Siluro-Devonian pelagic limestones.

grey and creamy limestones and dolomites with an east-dipping **anastomosing spaced foliation** and abundant brecciation.

The track descends to a T-junction by an ornamental pine tree: turn left here and descend further, passing abundant limestone exposures, to another metal fence viewpoint (at a hairpin bend in the path), this time looking north across the park (Fig. 16). From here the path descends south to rejoin Carrer de Castell-terçol. However, rather than enter the road, turn left past park benches (on the left) to enter a gate allowing access to the recreational area in the old quarry floor. Just beyond the cafeteria there are more good exposures of limestone on the right, close to the NE park exit and adjacent to the metal sculpture Tòtem (Location 3). These grey, creamy limestones contain abundant **diagenetic** brown crystals of oxidised sulphide and show a gently dipping, finely spaced anastomosing folia-tion affected by kink folds. Around the side of this exposure is a steeply dipping fault in which limestones and red shales have been converted into **cataclastic** fault rocks. The exact stratigraphic position of these deformed sediments is un-certain but they are tentatively attributed to the Olorda Formation of Julivert and Durán (1990a), with an age close to that of the Siluro-Devonian boundary (416 Ma).

Excursion 2: Les Turons

La Mare de Déu del Coll

The next exposure to be visited lies 400 m northeast, adjacent to the Romanesque church of La Mare de Déu del Coll. Exit the park descending NE from Tòtem to the road below (Passeig Mare de Déu del Coll) and turn left following signs to Carrer del Santuari. At this point we are crossing the normal fault which brings currently unexposed Mid-Carboniferous "Culm facies" sediments down against the limestones (Fig. 14). The more easily eroded nature of the Culm lithologies has produced lower ground between the limestone hills to west and east.

Walk 200 m uphill (north) and turn right (east) into Carrer del Santuari, which runs along the narrow ridge separating the Barcelona districts of Gràcia and Horta-Guinardó and provides a steep watershed between two distinct *riera* systems draining into the city from Collserola. Immediately on the right (Carrer del Beat Almató, opposite Meson Can Ramon) there are views across to the Llobregat delta above steps which descend steeply to the recently constructed metro station of El Coll-La Teixonera near the end of Line 5. The addition of this final segment to Line 5 (between Horta and Vall d'Hebron) caused a major social and political crisis in Barcelona when the tunnelling operations on the north side of the watershed led to a huge building collapse (in the barrio of El Carmel some 600 m NE from here). The collapse began on 25 January 2005 with the disappearance of a garage (at No. 12 Carrer del Calafell) into a hole 18 m wide and 35 m deep. This early warning prompted the evacuation of over 1,000 people in adjacent buildings and avoided any loss of life during a subsequent larger collapse (on 27 January) involving several apartment blocks. The tunnelling appears to have been taking place through the faulted Carboniferous sequence beneath the Turó del Carmel Thrust System.

Continue along Carrer del Santuari eastwards for 150 m to the church of La Mare de Déu del Coll, an 11th-century building extensively restored in the 20th century. There is a strong contrast between the use of mostly Palaeozoic sedimentary lithologies in the original walls and the untectonised brown Miocene sandstone utilised in the modern additions. Across the road, red conglomeratic sediments are exposed in a low cliff (Location 4, Fig. 17). These alluvial conglomerates, comprising an upward-fining, poorly sorted mixture of rounded-to-angular, mostly quartzitic **clasts** in a red **matrix**, are unaffected by Variscan deformation and interpreted (by comparison with exposures in other areas) as Lower Triassic in age. They presumably were deposited with angular unconformity upon the Palaeozoic sequence, although this unconformity is not exposed (and anyway may not be present here due to possible local involvement within the Turó del Carmel Thrust Zone).

The small exposures of Triassic rocks found here on the western side of the Turó del Carmel are the only post-Variscan/pre-Alpine sediments within Barcelona city limits. There is a larger, although still isolated, outlier of Triassic rocks behind the coast at Montgat, 8 km to the NE, and much more complete successions in the Garraf Massif immediately west of the river Llobregat and along

Figure 17. Coarse-grained, poorly sorted conglomeratic fluvial red beds belonging to the Triassic succession deposited unconformably upon a Palaeozoic basement deformed and metamorphosed during the Variscan orogeny. These are the only exposures of Triassic rocks in the Barcelona city limits and occur within a small, faulted outlier within the footwall of the Turó del Carmel Thrust Zone. They are underlain by Carboniferous "Culm facies" turbiditic sediments and overthrust by Siluro-Devonian limestones.

the NW side of El Vallés, behind the Collserola Hills (Fig. 2). These sediments belong to the widespread *Buntsandstein* facies deposited in continental **rift basins** running from Scandinavia to Iberia and beyond, during the early Mesozoic extensional break-up of Pangea (López-Gómez *et al.*, 2002).

Park Güell
To reach Park Güell, turn right at the church of La Mare de Déu del Coll, with the Triassic exposures now on the left, and walk south to reach immediately a triple road fork. Continue straight ahead (Carrer del Portell), staying horizontal, to pass an area with public seating on the right from which there are views back across El Coll to La Creueta del Coll and Collserola. (Note: this section of the route is currently (2011) undergoing extensive roadworks with the intention of laying a paved roadway towards the back entrance of Park Güell). Continue on the level along the same lane (which changes name from Carrer del Portell to Camí de Can Mora: Carrer del Portell forks off and descends to the right), curving to the left with open views across southwestern Barcelona. The lane contours the hill,

Excursion 2: Les Turons

following the main basal thrust which brings the Palaeozoic limestone sequence over the Triassic and Mid-Carboniferous sediments in the **footwall** (Fig. 14), and passes below the buildings of Escola Virolai which lie in the hanging wall above. Stay on this lane which soon reaches a short, sharp incline (with a wide track running off to the right) where the lane climbs onto Siluro-Devonian sediments in the thrust hanging wall and leads to the viewing platform of Mirador Virolai (Location 5). The 180° panorama from this popular viewpoint reveals the gently sloping Pleistocene platform (*Pla Alt*) on which much of the city is built. The faulted southeastern margin of Les Turons forms the northwestern boundary of the *Pla*. Directly southeast, beyond the grid of the 19th century enlargement to the city (L'Eixample) the modern coastline defines the southwest extension of the Besòs delta, to the right of which lie the shipping port and the hill of Montjuïc (the starting point for the next excursion). To the left the distant coastline runs along the southeast edge of the Littoral Range, the hills of which expose mainly Permian granitoid rocks of the Montnegre Massif (Fig. 2).

Return down the incline and take the wide track leading down to the left. This leads in turn to a signposted track junction with highly faulted, black-chert veined Silurian shales on the left (Fig. 18). Turn sharp right here following

Figure 18. Intensely deformed Silurian (?) veined dark cherty and contorted pelagic sediments exposed within the Turó del Carmel Thrust Zone below the Mirador Virolai (between Locations 5 and 6, Excursion 2).

the sign to Font de San Salvador. This track descends to another hairpin which curves left, passes Carrer de Pau Ferran on the right and continues down through the pine forest. Almost immediately there are poor exposures of thrust footwall Carboniferous sediments on the left: continue on a further 150 m or so to better exposures at another junction and go sharp left. Here on the left are relatively fresh exposures of these Mid-Carboniferous "Culm facies" turbiditic rocks which here comprise highly immature, dark micaceous sandstones and conglomerates disrupted by faulting and cut in places by arrays of quartz veins (Location 6).

Continue for another 150 m, curving right to a signposted junction and turn left to enter a cutting leading through a metal gate into Park Güell (Location 7). The cutting provides an excellent exposure of the tectonic contact between the Carboniferous siliciclastic rocks (footwall) and Siluro-Devonian calcareous rocks (hanging wall) juxtaposed by the Turó del Carmel Thrust Zone (Fig. 14, cross section C-D). The first 10 m through the cutting exposes grey, faulted Mid-Carboniferous sandstones which are then overlain by a 10-m-thick, southeast-dipping zone of intense shearing in which Palaeozoic protolith lithologies have been reduced to fault rocks. Most of this zone comprises alternations of calcareous breccias, faulted shales (similar to those seen in Fig. 18) and thin, gouge layers. Adjacent to the metal entrance gate these fault rocks are overlain by calcareous rocks attributable to the pelagic limestone succession. Against the fault zone these too are severely affected by intense shearing and folding, with red shaly and paler calcareous finely foliated lithologies wrapping around less-deformed areas of limestone. In the final few metres of exposure there are vertically bedded and folded pelagic limestones which sharply abut the fault zone. Given the complexity of fault fabrics through this section, and the tectonic setting of the Turons Fault Complex, it is likely that there have been repeated reactivations, extensional as well as compressional, along this prominent movement zone.

Enter the park (in 2011 still free of charge) to enjoy more views over the city, then turn left and follow the track past Casa Tríes then more abundant limestone exposures, down to a series of broad curves, first right (Lovers' Bridge and Hanging Gardens) then curving left to reach exposures of tightly folded, thinly bedded Silurian black cherts and mudstones in cuttings on the left, below the Lovers' Bridge (Fig. 19). After this, the final curve to the right leads to the track feeding in from the eastern park entrance. Here turn right and follow the horizontal track straight on southwest for nearly 200 m to the (daytime only) Bar de la Cueva and the central viewing platform over the city (left). On the right of the track here are more exposures of cherty Silurian pelagic sediments below Gaudí's curving ornamental wall. The restrooms here are excavated into the rock face and provide an unusual opportunity for further examination of the Silurian sequence. If this is your first visit to Park Güell, then allow an extra hour to explore this popular tourist attraction, concentrating on the lower slopes of the park.

From the restrooms behind the Bar de la Cueva, continue following the same track (Passeig de les Palmeres) which curves left, right, then left again to

Figure 19. Sub-isoclinally folded, thinly bedded Silurian black cherts and mudstones exposed within Park Güell. Most of this park, which is one of the most tourist-visited sites in Barcelona, lies on the Siluro-Devonian pelagic shale-chert-limestone sequence in the hanging wall of the Turó del Carmel Thrust Zone. Excursion 2, Location 8 (Fig. 14).

reach another metal entrance gate, this time in the southwest corner of the park. Ahead, forming the cliffs along the right side of the road (Carrer Sant Josep de la Muntanya), are more exposures of the same Silurian sequence (Location 9). To the left are the steps figured on the front cover, leading down to the busy main entrance to Park Güell. Descending these steps one can enjoy how Gaudí has integrated exposures of the faulted metasedimentary basement rocks into the architectural form of the staircase and adjacent walls (built from blocks of local limestone). In the three blind windows in the wall at the base of the steps the contact is exposed between Palaeozoic basement and overlying Recent cover of poorly sorted, locally derived hillwash breccias. Climbing back to the top of the steps, there are views northeast over to another prominent hill, Turó de la Rovira which, at 261 m, is only slightly lower than Turó del Carmel to the left. Turó de la Rovira once again provides extensive exposures of the Palaeozoic limestone sequence, whereas in the low ground in front (the SE corner of Park Güell) there are poor exposures defining an inlier of the Cambro-Ordovician sequence seen at the beginning of the excursion (Fig. 14). Re-entering the park back through the Carrer Sant Josep de la Muntanya gate, climb immediately left up the track which snakes back up-stratigraphy to a junction where a track leads left to another park exit (Baixada de la Glòria). This exit descends metal steps to the urbanised hillside of Baixada de la Glòria (signposted Av. de Vallcarca and Metro). Descend the steep hillside using several flights of steps with views ahead across the valley to Putget. At the valley bottom, ignore signs for Vallcarca Metro to the right and instead cross the main road and continue straight on, climbing up Carrer d'Agramunt for a short distance to the next crossroads to turn right into Carrer de Bolivar. After climbing a further 150 m this road merges into Avinguda de la República Argentina where we continue in the same direction to reach the Vallcarca metro entrance directly ahead. Immediately before this metro entrance, cross the road at the traffic lights (left) to reach the final exposure of this excursion: a roadcut with explanatory geological panels demonstrating another thrust zone. This one has Silurian black slaty siliceous mudstones in the hanging wall and Siluro-Devonian limestones in the footwall (Location 10). This thrust, which we suggest belongs within a southwestward continuation of the Turó del Carmel Thrust Zone, lies close to the underlying younger Creueta Fault into which it appears to merge (Fig. 14). Back across the road at the Metro entrance there are views back across to Escola Virolai and Turó del Carmel.

Excursion 3: Montjuïc to the Medieval City

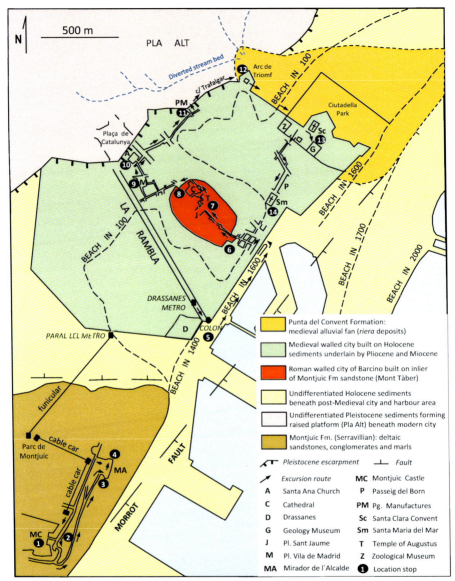

Figure 20. Route map for Excursion 3 which begins on the Miocene rocks of Montjuïc then descends to the modern harbour front, traverses the Roman core of the city (Barcino) and continues on to reach the Pleistocene Platform on which much of the post-Medieval city has been built. The last part of the excursion follows the margin of the Pleistocene escarpment then descends the outcrop of the Punta del Convent Formation (alluvial fan) to the Geology Museum in Ciutadella Park, before re-entering the Medieval city to reach the church of Santa Maria del Mar via the Passeig del Born. The excursion explores the inextricable linkage between geology, archaeology, history and culture, as illustrated by the mixture of geological and historical boundaries on the map.

EXCURSION 3: MONTJUÏC TO THE MEDIEVAL CITY

This excursion focuses on the "young geology" of Barcelona, beginning on the hilltop of Montjuïc (192 m) which overlooks the city and where Miocene sediments are exposed, then descending to the modern sea front and Medieval city (Fig. 20). Given the intensely urbanised nature of this terrain, there are inevitably few exposures, although there is much of geomorphological and historical interest. The route passes many of the city's attractions, including Montjuïc (castle, viewpoints, museums), Las Ramblas and the harbour front, the Roman heart of the Medieval city (Roman walls, Plaça de Sant Jaume, Forum, Marés Museum, cathedral, necropolis), the modernist architectural masterpiece of the Palau de la Música Catalana, Ciutadela Park, Geology Museum, Passeig del Born, and finishes inside the magnificent 15th century church of Santa Maria del Mar. A full day (preferably Tuesday to Friday) is recommended.

Montjuïc

To reach Montjuïc Castle take the Metro green line (line 3) to Paral.lel and transfer to the funicular railway which ascends one stop to the Parc de Montjuïc terminal (90 m). The funicular exits into a road with a bus stop immediately on the left and a cable car station to the right. Either take bus 193 (you may have to wait a while) or the cable car (note: T10 not usable on cable car): both climb nearly 100 m to a stop at the steps below the castle. Follow these steps to the castle entrance (currently free entry: 2011), which provides a fine example of the use of the brown sandstones that characterise much of the Miocene strata forming the hill (Fig. 21, Location 1). Inside the castle from the defensive walls there are impressive views across the city and its surroundings.

The mid-Miocene sandstones of the Montjuïc Formation have been utilised as masonry stone at least since Roman times, and are ubiquitously present in buildings throughout the Medieval city. The exceptionally quality of much of the stone, which is strong, compact and durable yet easy to work with, owes its origin to pervasive diagenetic alteration of the sandstone during **lithification** (Gómez-Gras *et al.*, 2001). The early growth of a hard cement of **authigenic** silica and K-feldspar has given the best of these sandstones (known locally by quarrymen as *el blanquet*) their massive, homogeneous character. Most of the larger quarries excavated into Montjuïc lie on the west and northwest side of the hill, within the Castell Member. During the first half of the 20th century there were around 25 active quarries. The industry was closed in 1955 and many of the quarry faces are now variously overgrown and difficult to access. The Teatre Grec is a notable exception, where the quarry face is used as a backdrop to an open air stage.

The structure of Montjuïc is that of a tilted fault block dipping northwestwards and bounded abruptly on its seaward side by a steep, southeast-dipping normal fault (Morrot Fault: Fig. 20). To view the escarpment created by erosion along the Morrot Fault, walk towards the sea from the castle entrance to locate

Excursion 3: Montjuïc to the Medieval City

Figure 21. Use of high quality sandstone in the front entrance to Montjuïc Castle (Excursion 3, Location 1). The early growth of a hard cement of authigenic silica and K-feldspar has given the best of these sandstones their massive, homogeneous character. The present castle was constructed in 1751 on the remnants of an earlier one built in 1640. Before these, buildings on the hillside utilising the local stone included a 14th century watchtower, various chapels (the oldest dating back at least to the 10th century) and remains left by Romans and Iberians.

steps on the left. These descend to the path that runs beneath the castle and overlooks the cliffs which drop to the port 170 m below. To the right there are extensive views over the mouth of the Llobregat delta.

Although the Morrot Fault is one of the most visually apparent of the extensional structures that influence the landscape of Barcelona, geologically it is not the most important. It appears to be a relatively minor splay from a much larger structure, the Barcelona Fault, the main trace of which runs parallel to the shoreline on the seaward side of the port (see cross section on Fig. 2). Geophysical data image the Barcelona Fault as having a listric geometry curving southeastward down to a mid-crustal **detachment surface** lying at 12–16 km beneath the Gulf of Valencia (Roca & Guimerà, 1992; Roca *et al*., 1999) and it has been estimated to have extended the crust by around 11 km during Miocene times (Gaspar-Escribano *et al*., 2002). Initial Miocene extension along the Barcelona Fault was rapid, although by Mid-Miocene times, when the Montjuïc sediments were being deposited, movement had decelerated and sedimentary infilling of the basin created by the previous extension was taking place (Gaspar-Escribano *et al.*, 2002). Like other faults in the area, the system of normal faults linked into the main Barcelona Fault has had a long history of polyphase movement, which is mostly of Miocene age but with displacements appearing to affect sediments as young as Pleistocene (Liquete *et al.*, 2007; Gámez *et al.*, 2009).

Follow the clifftop path left (north) through pine forest with views east over the city and the Costa Brava beyond. At the second lamppost there are exposures of Montjuïc Formation sediments on the left (Location 2, Fig. 20). The siliciclastic sedimentary succession exposed on Montjuïc was deposited around 12–13 Ma (Serravillian) in a Miocene fan delta where rivers draining the ancestral Collserola Hills disgorged into the sea. These sediments are mostly sandstones, although there are also coarser (conglomeratic: Fig. 22) and finer (marl) lithologies. Overall, the sequence has been divided into four main lithostratigraphic units (Gómez-Gras *et al.*, 2001 and references therein), which are classified here as Members within a Montjuïc Formation. These four Members (and their approximate thicknesses) are, from base to top, Morrot (>80 m), Castell (100 m), Miramar (15 m) and Mirador (>20 m). The sedimentary environments

Figure 22. Conglomeratic masonry block utilised in the wall of Montjuïc Castle. Conglomerates within the Miocene Montjuïc Formation are interpreted as channel deposits within a delta plain environment (Gómez-Gras *et al.*, 2001). Clasts within the conglomerate are derived mainly from the Palaeozoic basement exposed in the nearby Catalan Coastal Ranges (Collserola Hills). The conglomerate block in the wall is surrounded by blocks of finer sandstones more typical of the Montjuïc Formation and which were used to build most of the Medieval city of Barcelona.

under which these deposits formed ranged from mixed fluvial-marine delta plain to dominantly marine conditions at the delta front and beyond (Gómez-Gras *et al.*, 2001).

The northwesterly dipping sediments exposed here beside the path belong within the Castell Member and comprise an alternating series of sandstones and conglomerates (Fig. 23) interpreted as channel deposits in the delta plain environment (Gómez-Gras *et al.*, 2001). The pebbly conglomerates, which show ferruginous, red-stained joint surfaces, fine up from an erosive base and contain a variety of clasts that include white vein-quartz, pale weathered granitoids, dark slaty metamorphics and fossils (mostly bivalve fragments). Most of the coarser clasts in these rocks were derived from erosion of the ancestral Collserola Hills metamorphic and granitoid basement and were subsequently washed down by *riadas* into the marine delta environment and mixed with finer sands and shell debris. The finer, sometimes bioturbated, sandstones include softer, poorly cemented varieties as well as layers of the hard, brown *blanquet* type preferred for building stone.

Figure 23. Miocene sandstones and conglomerates of the Castell Member (Montjuïc Formation) exposed in the path below Montjuïc Castle (Excursion 3, Location 2). The sediments dip gently NW in the footwall of the Morrot Fault (see text for details) and TM is pointing to a prominent conglomeratic layer within which can be found fossil fragments, rounded vein quartz pebbles and both metamorphic and igneous materials washed from the ancestral Collserola Hills into the Mid-Miocene marine delta.

Continue on past these exposures and a low building (Caseta Jardineres), descending and staying to the right. There are no exposures as the path crosses the overlying soft marls that comprise the Miramar Member and descends more steeply eventually to pass a fountain on the left and leads (right) down to the Mirador de l'Alcalde. This upper viewing platform, lying at 126 m, offers views southwest back across the side of Montjuïc Hill and over the port (Location 3, Fig. 20). There are exposures of soft, pale grey Miramar Member marls underlain by Castell Member conglomeratic sediments in the cliff exposures ahead of and below the platform. Now descend northeast to the lower viewing platform which affords views over the city and the hills beyond: Les Turons, the Collserola Hills and their northeast continuation towards the Costa Brava (Location 4, Fig. 20).

From here it is enlightening to consider the Pleistocene-Holocene geology underlying the city between Les Turons and the sea. Behind La Sagrada Familia can be discerned the seaward-sloping, interfluvial piedmont surface that fashions much of the topography between the Besòs and Llobregat deltas. The sediments underneath this slope are mostly Pleistocene hillwash sands and gravels, channelised flow deposits, windblown (loessic) silts, and calcareous *caliche*-type mudstones and **palaeosols** which together form the *Pla Alt* or Higher Platform on which, for example, the gridplan 19th century city enlargement (l'Eixample) was built. In contrast, the Medieval city below (and all the ground seaward to the right of the prominent Agbar Tower) rests mainly on overlying Holocene sediments and lies several metres below the *Pla Alt*. An exception is provided by the small hill around the Cathedral (rising to only 17 m above sea level and so imperceptibly low from this distance), known to the Romans as *Mons Taber* and the site they chose to found the walled city of Barcino (Fig. 20).

To the right of the ancient city centre, the coastline has progressively prograded seaward since Roman times to reach its present position around Port Olímpic (Fig. 24), as demonstrated by the position of the former beach strands outlined on Figure 20. Part of this progradation process was the rapid growth of an alluvial fan on the east side of the walled Medieval city, a result of the diversion of several *riera* watercourses around the fortified urban centre. This anthropogenically enhanced alluvial fan, known as the Punta del Convent Formation (Riba & Colombo, 2009 and references therein) after the 13th century Santa Clara Convent built in what is now Ciutadella Park, is full of coarse debris washed down from the Palaeozoic rocks of Les Turons. The trace of its outcrop, raised up to 5 m above the adjacent land surface, runs 2 km eastward from the Arc de Triomf down to the coast at Port Olímpic, through Ciutadella Park (Fig. 20). An unpredicted outcome of the aggradation of this alluvial fan during the 250 years after the eastern Medieval walls were erected was that the raised level of the ground against the wall facilitated forced entry into the city during the siege of 1714 when Catalan forces were overcome by French and Castillian troops on 11 September (now commemorated as the National Day of Catalonia). The green strip of Ciutadella Park, where the storming of the Medieval walls from the al-

Excursion 3: Montjuïc to the Medieval City

luvial fan took place, broadly follows the outcrop of this alluvial fan and can be discerned from this viewpoint (Fig. 24): it will be visited later in the excursion. From Ciutadella Park the main road can be seen running south to the 60-m-high column of the Colon Monument, rising below from the seaward end of Las Ramblas (Fig. 24, Location 5 on Fig. 20). The next part of the excursion begins outside Drassanes Metro station at the foot of this column.

Figure 24. View looking NE from the lower Mirador de l'Alcalde over the Barcelona coastline (Excursion 3, Location 4). The Colon Monument (arrow) rises from what was once the beach lying alongside the SE corner of the old city. Behind the monument the wide Passeig de Colon runs towards Parc Ciutadella, which lies on the alluvial fan of the Punta del Convent Formation at the former NE corner of the fortified Medieval city. To the right of the old city, the coastline has progressively prograded seaward since Roman times to reach its present position around Port Olímpic, as demonstrated by the position of the former beach strands outlined on Figure 20. In the distance the granodioritic hills of the Montnegre Massif (Littoral Range) run out to meet the coast at the Costa Brava.

Modern Coastline to Barcino

To reach the Colon Monument, leave the viewpoint, passing a refreshments kiosk (on the right) and go back to the road. Up on the left here (up steps) is a wave-ripple-marked block of ferruginous Montjuïc sandstone bearing the title *Mirador de l'Alcalde*. The entire Mirador area rests on the Mirador Member, the youngest Miocene unit exposed within the Montjuïc Formation, but *in situ* exposures have been lost. To reach Drassanes Metro station from here, there are various options. That involving least effort is to take the infrequent 193 bus for an entertaining ride down Montjuïc to Plaça Espanya, then connect with the metro (green) line 3 to Drassanes. Alternatively, to avoid waiting for a bus, walk back to the Parc de Montjuïc funicular terminal by crossing the road, turning right to pass the Plaça de la Sardana sculpture (fossiliferous limestone) and continue down the road to locate the small park of Jardins de Joan Brossa. Look at the map displayed at the park entrance to locate the Plaça de Dante on the opposite side of the park. There are various routes through the park to reach this Plaça, the easiest being to take the path immediately down to the right (past *Pinus pinaster* sign) then first left to reach a water drinking fountain and steps leading down to a wooden boardwalk. The boardwalk leads on into a gravel path (with the road below on the right), passing an Olympic sculpture in memory of Joaquim Blume, and eventually connects with a paved path that leads down to Plaça de Dante. Walk left around Plaça, keeping the fountain on the right, to locate the funicular station and metro connection once again. From here, either return to Paral.lel station and take the metro one stop to Drassanes, or if time allows, widen the exploration of what Montjuïc has to offer by continuing along the road to pass the Miró Museum and Olympic Stadium and then descend via the excellent MNAC (National Museum of Catalonian Art: there are coffee bars both outside and inside the Museum, no need to pay the entrance fee) and the Magic Fountain to Plaça Espanya (then metro to Drassanes). Those with a map and sense of direction can make a diversion to the Teatre Grec, in the gardens below the Miró Museum, where the backdrop to the open stage is an old quarry face preserving a good exposure of Castell Member sediments.

The Drassanes metro exit emerges on the seaward end of Las Ramblas, where the *Riera d'en Malla* originally disgorged its sedimentary load eroded from Collserola and Les Turons alongside a 13th century city wall ending at the sea. This *riera* was diverted northeast around the city in 1447 to allow construction of a new wall crossing the river at Canaletes (the entrance to what is now the Plaça de Catalunya) and the dry course of the abandoned stream deposits became a new focus for the city, eventually evolving to Las Ramblas of today (Hughes 1992; Arranz Herrero, 2003; Riba & Colombo, 2009).

Walk out to the seafront, crossing traffic lights and passing the Colon Monument to reach Las Golondrinos boat terminus (Location 5). Looking back inland from the sea front, to the left rises Montjuïc, the distinctive profile of which is defined by the fault-bounded **scarp slope** to the southeast and the more gentle

dip slope to the northwest. Ahead, between Montjuïc and Las Ramblas lies the relatively low-lying area of Raval which was previously a marshy area protected behind a coastal sand ridge and being fed by *rieras* from Collserola. Holocene sediments underlying this wetland date back to at least 7,000 BP and during the early growth of the city the area was infamous for its unhealthy character. Even since draining and urbanisation, this part of town is subject to flooding during times of exceptional rainfall. The nearby medieval naval arsenal of Drassanes (now the Maritime Museum, across the main road left of Las Ramblas) was built at the seaward end of these marshes, just behind a protective sand barrier which ran from below the Montjuïc cliffs northeast to what is now Ciutadella Park.

Walk northeast along the harbour front for 600 m, with the sea on the right, passing the Rambla del Mar footbridge and following the line of the beach once lying in front of the now demolished medieval wall (La Muralla de Mar) at the turn of the 17th century. Around 100 m after where the waterfront ends, head left inland to cross the busy road system at multiple traffic lights, passing a colourful sculpture (right) to reach the Passeig de Colom. Turn left here for a short distance then right to enter the Medieval city (Fig. 20) via the narrow alleyway of

Figure 25. The four remaining columns of the 1st century BC Temple of Augustus in what was once the forum of Barcino, built on a low hill (17 m) of Miocene sandstone adjacent to the coast (Excursion 3, Location 7). The forum was the political, economic, administrative and religious centre of the Roman city.

Carrer de Marquet. Follow this narrow lane which climbs steadily up towards the Roman core of the city, passes three crossroads, and en route changes its name to Carrer de Regomir. Just beyond the junction with Carrer del Correo Viejo we enter what was once the Roman city. The seaward entrance gate to the Roman city (*Puerta Decumana al Mar*) originally stood adjacent to what is now the Centro Civic Pati Llimona (No. 3 Carrer de Regomir: Location 6 on Fig. 20). The Roman remains in this medieval building, which include a part of the original walls of the city founded in 15 BC, can be viewed inside this Civic Centre.

Barcino to the Pleistocene escarpment

From the Civic Centre continue on up Carrer de Regomir, which becomes Carrer de la Ciutat and emerges in Plaça de Sant Jaume, the political heart of Catalonia, with the city hall immediately on the left (Ajuntament) and the Parliament (Generalitat) opposite. Cross the square, heading slightly right, to enter the narrow street of Carrer de Paradís in the far corner. Still gaining elevation, this street turns abruptly right to reach the summit of Mont Tàber at the next corner, marked by a wall plaque (16.9 m) on the right. Here, a doorway allows free entrance to view remains of the 1st century BC Temple of Augustus (Fig. 25; Location 7 on Fig. 20) in what was once the Roman city forum, built on top of the unexposed Miocene sandstone inlier that underlies the hill.

Return to the street and continue to the end of Carrer de Paradís: ahead is the back of the Cathedral. Keeping the Cathedral on the left, walk around it to enter Carrer dels Comptes. In this street is the entrance to the Museo Marés (Plaça de Sant Iu), formed by the portal of the relocated Santa Clara Convent, the original convent having been destroyed as a consequence of the 1714 invasion of the city as mentioned earlier (Figs 26a, b). Continue on Carrer dels Comptes past the museum and on to enter the square in front of the Cathedral: this is now the other extremity of Roman Barcino. Turn left, curving past the Cathedral front entrance to view remains of the new Roman wall built under the orders of Claudius II around 270–310 AD, in response to raids by Germanic tribes and replacing the original, poorer quality wall. This superbly constructed defensive structure, built from Montjuïc sandstone, included 78 towers, several of which are preserved here. On the far corner, in Plaça Nova at the entrance to Carrer del Bisbe, was the northwestern entrance to the Roman city (*Puerta Praetoria de Montaña*: Fig. 27; Location 8 on Fig. 20).

The next point of interest is the site of a Roman necropolis preserved in Plaça de la Vila de Madrid, which lies a short distance west of the Cathedral. With Carrer del Bisbe and the Roman wall behind, fork left into Carrer dels Boters (not Carrer de la Palla) and after 100 m pass Plaça de la Cucurulla and turn right into Carrer del Duc. Once in Carrer del Duc, take the first left into Carrer de Francesc Pujols which leads into the Plaça de la Vila de Madrid (Location 9). The necropolis here has over 90 known graves which lay on either side of one of the roads leading into the city. Many tombs are marked by semi-cylindrical monolithic fu-

a

b

neral monuments called *cupae*, usually carved from Montjuïc sandstone and very distinctive to Barcino (Bonneville, 1981), which can be viewed from the square (Fig. 28) or examined inside the small museum here. The Roman remains rest on (now unexposed) alluvial sediments containing clasts of Palaeozoic granitoid and metasedimentary rocks as well as pre-Roman (Iberian) ceramic fragments. The Roman stratigraphic level, lying 1–2 m below modern ground level, is overlain by mixed anthropogenic (landfill of building foundations) and natural (alluvial sediments) materials. The sediments include sand layers presumably deposited during flood events from the nearby *Riera d'en Malla* in what is now Las Ramblas (Riba & Colombo, 2009).

Figure 27. Roman use of Montjuïc Formation Miocene sandstone in the *Puerta Praetoria de Montaña*, once the NW entrance to the Roman city and now the NW corner of the medieval Cathedral (Excursion 3, Location 8).

Figure 26. (Facing page) The portal to the Frederic Marès Museum (Fig. 26a) adjacent to the Cathedral originally stood as the entrance to Santa Clara Convent in the Plaça del Rei (Palau Reial Major). The ill-fated original Santa Clara Convent was built in what is now Ciutadella Park on the outcrop of the fluvial Punta del Convent Formation. Continued fluvial sediment aggradation against the city walls during Medieval times rendered the convent especially vulnerable to attack during siege, leading to repeated abandonment and damage in times of war. Finally, following the surrender of Barcelona to Spanish and French troops in 1714, Felipe V transferred the religious community from what was left of the old convent ruins to Tinell in the Palau Reial Major and a new convent was established. Fig. 26b shows the portal to the new convent in its original site in the Plaça del Rei (reproduced by kind permission of the Frederic Marès Museum).

Figure 28. Use of Miocene Montjuïc Formation sandstone in the distinctive semi-cylindrical monolithic funeral monuments called *cupae* discovered within the Roman necropolis preserved beneath the Medieval city at Plaça de la Vila de Madrid (Excursion 3, Location 9).

Exit from the Plaça in the far opposite corner to the *cupae* and enter Carrer de Bertrellans which leads to Calle de Santa Ana. Turn right and immediately on the left is a gate access to Plaça de Ramon Amadeu and the 12th century church of Santa Ana (Location 10). The 15th century cloister (open mornings only) adjacent to the church is geomorphologically interesting because it was built at an angle to the church against the low escarpment of the Pleistocene Platform (*Pla Alt*) that runs along the northwestern boundary of the Medieval city. The adjacent Plaça de Catalunya lies on the *Pla Alt* and so is 3–4m higher than the cloister. Thus, the car park here at ground level (left side of the Plaça de Ramon Amadeu) is entered as an apparent "basement" car park from a higher level off the Plaça de Catalunya. An appreciation of this difference in height is gained by ascending the short flight of steps which climbs to the car park roof (now the Plaça de Ramon Amadeu) on the far side of which a gate provides access into Carrer de Rivadeneyra and Plaça de Catalunya. The urbanisation of this geomorphological feature, known in Catalan as *el graó barceloní*, has left only subtle evidence for its former existence, but it can be traced from Plaça de la Universitat to Plaça de Catalunya (compare the ground levels in Carrer de Pelai and Carrer dels Tallers), and on past Santa Ana Church to the northern edge of the old city.

To visit another curious, urbanised remnant of this Pleistocene escarpment, return through the entrance gate to Plaça de Ramon Amadeu, turn left (northeast) back into Calle de Santa Ana at the end of which cross Avinguda del Portal de l'Àngel to enter Carrer de Comtal, which leads in turn to the traffic congested Via Laietana. Keeping right, cross Via Laietana at traffic lights and dogleg right-left to continue in the same direction along Carrer de Sant Pere Més Alt. The spectacular Palau de la Música Catalana is on the left, the bar and restrooms inside at the back of which make an excellent comfort stop (access around the left side of the building). Back outside the Palau, continue along Carrer de Sant Pere Més Alt for another 100 m to locate black metal gates on the left at No. 31/33. When open, these gates allow public access to a narrow, tiled and roofed passageway known as the Passatge de les Manufactures. Inside the passage, pass the Bar Pasajes and right at the back there is a flight of 26 steps that climbs the *graó* escarpment and so provides access to the Pleistocene Platform on which L'Eixample, the 19th century enlargement to the old city was built (Location 11). (Note: if the passage is closed, then continue along Carrer de Sant Pere Més Alt to take the first left (Carrer de Méndez Nuñez) then first right into Carrer de Trafalgar and on to the Arc de Triomf).

Punta del Convent Formation to Santa Maria del Mar
Climbing the steps to emerge from the Passatge de les Manufactures, turn right into Carrer de Trafalgar and follow it for 400 m to the Arc de Triomf. This is now at the head of the previously mentioned Punta del Convent Formation alluvial fan (Riba & Colombo, 2009), enhanced by the diversion of *riera* floodwaters around the Medieval walled city (Fig. 20). Roadworks adjacent to the Arc de Triomf in 2010 exposed several metres of crudely stratified, reddish-brown alluvial breccio-conglomerates and sands (http://blocdecamp.blogspot.com/2010/03/la-formacio-punta-del-convent-les-obres.html). In this area the *graó* escarpment has been entirely lost by the combined effects of *riera* activity and subsequent urbanisation. Beyond the Arc de Triomf area, however, the escarpment can be traced curving northwards to the River Besòs and the eastern edge of the Collserola Hills (Fig. 2), gradually rising in height to nearly 10 m.

Standing under the Arc de Triomf (Location 12) the view inland follows the line of the broad Passeig de Sant Joan which gently climbs the Pleistocene Platform. In the distance are the slopes of Park Güell (Excursion 2) behind which rise the Collserola Hills (Excursion 1). Pass through the Arc de Triomf and stroll seaward down the wide Passeig Lluis Companys towards the sea, following the outcrop of the Punta del Convent Formation to reach Ciutadella Park. Continue into the park, where lies the former site of the 13th century Santa Clara Convent (Fig. 20). Nothing remains of the convent *in situ*, although various fragments of the building were subsequently dispersed around the city, with pieces of the gothic cloister for example ending up in the Marés Museum passed earlier in this excursion (Fig. 26). The Zoology and Geology museums occupy buildings to the right, separated by the L'Hivernacle glasshouse. Although both museums are currently closed (2011) they are due to re-open after renovation for a major exhibition. Choosing a seat in the park or the Geology Museum if open (Location 13), read the following section before moving on.

Given the emphasis placed during this excursion on geologically very young events, it is interesting finally to consider the neotectonic setting of the city. As will already be clear from the localities previously visited, the geology of Barcelona is strongly influenced by faulting. Indeed the whole city area lies within what is essentially a link zone between the Barcelona Fault and the Littoral Range of the Catalan Coastal Ranges (Fig. 2). Although the main displacements along the major faults took place relatively long ago during Palaeogene and Early Neogene compression and extension, the presence of fault-bounded Mid-Miocene marine sediments at an altitude of nearly 200 m on Montjuïc indicates younger seismic events have occurred in the Barcelona area. Furthermore, some of the faults proved by borehole studies beneath the city appear to displace not only Pliocene and Pleistocene age strata, but also Holocene sediments (Riba & Colombo, 2009 and references to reports therein). It would appear from earthquake focal mechanism studies that the extensional event associated with Neogene rifting of the Valencia Trough has passed and that northeast Spain is currently once again under a N-S compressive regime (Olivera *et al.*, 1992).

In addition to the NE-SW fault set that dominates the Catalonian coastal zone (Penitents, Morrot, Barcelona, etc.), there are also NW-SE oriented transverse structures that have influenced, for example, the position of the Llobregat river delta, Mont Tàber and the SE boundary of Montjuïc. These transverse faults in some cases appear to be only short linking faults connecting larger NE-SW structures, but others show significant recent movements, and the Llobregat Fault Zone runs seaward into a major submarine fracture zone associated with volcanism (Maillard & Mauffret, 1999, Fig. 1). A high-precision levelling study undertaken in Catalonia by Jiménez *et al.* (1996) implicated displacements along these NW-SE cross faults as being not only responsible for measurable vertical movements during the 20th century but also linked to recent seismicity.

Current seismic risk for Catalonia is considered to be moderate (Liquete *et al.*, 2007), with major historical earthquakes including those of the "Great Seismic Crisis" of 1427–1428 (Olivera *et al.*, 1991, 2006; Figueras *et al.*, 2006). This crisis, centred on a NW-SE zone running between Amer and Queralbs in the Pyrenees 100 km north of Barcelona, destroyed several villages in the area and produced widespread damage in Catalonia. The most infamous seismic incident in Barcelona at this time occurred at the western entrance to the church of Santa Maria del Mar on the morning of 2 February 1428 where 25 people were reportedly killed by the partial collapse of the large *rosetón* window above the front entrance (Banda & Correig, 1984; Olivera *et al.*, 2006; Murcia, 2008). The epicentre of this destructive earthquake was at Queralbs (Banda & Correig, 1984), where most of the village population died. Twenty years later another big earthquake (estimated as Intensity VIII) occurred on the night of 24 May in the Vallès Oriental on the NW side of the Montnegre Massif with an epicentre at Llinars del Vallès, some 35 km NE of Barcelona (Salicrú i Lluch, 1995). This event also caused deaths and considerable damage in the region, notably in the Mataró area along the coast from the city, and to the castle at Papiol, above the river Llobregat on the west side of the Collserola Hills. This was the last of a series of damaging Medieval earthquakes affecting Catalonia, which today continues to be relatively seismically inactive.

With these events in mind, this excursion concludes with a visit to the church of Santa Maria del Mar. Leave Ciutadella Park by the exit adjacent to the L'Hivernacle glasshouse between the Geology and Zoology museums, cross the main road and continue down Carrer de la Princesa. Take the first left into Carrer del Comerç and continue past Carrer de la Fusina to reach the wide, tree-lined Passeig del Born on the right (Fig. 20). At the far end of this Passeig, in one of the most attractive parts of the Medieval city, stands the magnificent church of Santa Maria del Mar (Location 14). The back of the church, built from Montjuïc sandstone (like almost everything else in this area) can be accessed from the Passeig del Born: enter from the back or side door to the left. Robert Hughes (1992) has commented: 'There is no grander or more solemn architectural space in Spain than Santa Maria del Mar', and the effect of a visit here is always visually stunning. High above the front entrance to the church is the enormous *rosetón* window, replaced in 1459 after the earthquake damage of 1428 (Fig. 29).

The excursion ends here, in an earthquake-damaged Medieval church built on a former Roman shoreline where beach sands were laid on Plio-Quaternary sediments resting above an older sedimentary succession produced in response to Neogene extension during the opening of the Valencia Trough. As the African Plate continues on its long collisional course with Eurasia, the Alpine Orogeny is clearly far from over and we can confidently predict that Barcelona will remain geologically interesting for many more years to come.

Figure 29. The large *rosetón* window above the front (west) entrance to the church of Santa Maria del Mar (Excursion 3, Location 14) was partially destroyed during an earthquake on the morning of 2nd February 1428, killing a group of people gathered below. The window was restored in 1459. The earthquake was one of several in the "Great Seismic Crisis" of 1427–1428 and the probable cause of cracks in the façade. A previous earthquake during the construction of the church had earlier caused the collapse of the eastern tower in 1373.

GLOSSARY

Anastomosing spaced foliation: foliation in a rock is a non-genetic term meaning a planar structure such as bedding (sedimentary rocks) or cleavage (low-grade metamorphic rocks such as slates). In metamorphic rocks when the foliation planes are separated by 10 microns or more, they are referred to as spaced. Anastomosing means a branching, wavy, interconnected network, as in streams or leaf veins.

Antiform: whereas an anticline is an arching fold with the oldest rocks in the core, an antiform has the same shape but with no implication as to the relative ages of the rocks involved. The term is therefore commonly applied in folded metamorphic rocks where there is no fossil or sedimentological evidence for younging direction.

Aplitic/Pegmatitic: aplites and pegmatites are common in many granitic complexes, typically crystallizing rapidly at a late stage when most of the pluton has solidified. Aplites are pale, quartzo-feldspathic, fine grained and commonly sugary in texture, whereas pegmatites have very coarse crystals of quartz, feldspar and mica, each measuring several centimetres to over a metre in extreme cases. High levels of normally rare elements (such as boron and lithium) concentrate in late-stage magmas, so that pegmatites may contain unusual minerals such as tourmaline, topaz or lepidolite.

Authigenic: an authigenic mineral in a sedimentary rock was formed where it is found, crystallizing during diagenetic changes after deposition.

Axial planar fabric: in a simple upright fold such as that in Figure 7 a vertical central plane can be drawn through the middle between the two symmetrical sides (limbs) of the fold. This is the axial plane or axial surface, containing the fold axis and being broadly parallel to the metamorphic foliation.

Batholithic: a batholith (Greek: *bathos* = depth; *lithos* = rock) is a large mass of igneous rock intruded at depth (plutonic intrusion) in the Earth's crust and crystallized slowly, commonly heating the surrounding solid rocks enough to induce the growth of new minerals, a process referred to as contact metamorphism.

Breccia: a rock comprising angular rock or mineral fragments. Sedimentary breccias imply less sedimentary transport than conglomerates (which have more rounded fragments) and are typically mixtures of different clast sizes, a texture referred to as being "poorly sorted" and "sedimentologically immature".

Calc-alkaline: there are two main types of magma series that together produce most of the igneous rocks in the Earth's crust as they fractionate from mafic (rich in heavier elements e.g. iron and magnesium) to **felsic** (rich in lighter elements e.g. silicon and aluminium) compositions. These are the tholeiitic and calc-alkaline magma series. An important difference between them is that calc-alkaline magmas are more oxidized leading to the early crystallization of magnetite, which inhibits iron-enrichment relative to magnesium (unlike in tholeiitic magmas). Instead, during fractional crystallization iron and magnesium reduce

Glossary

in similar amounts (relative to aluminium) as calc-alkaline magmas evolve from gabbros through diorites and granodiorites to granites.

Calc-silicate: during metamorphism, a rock composition rich in both calcium carbonate and aluminium silicate, such as a muddy limestone, will produce a range of new minerals that also contain calcium and silica, such as the calcium garnet grossular ($Ca_3Al_2Si_3O_{12}$).

Caliche: a hardened, cemented sedimentary rock underlying the surfaces of many arid or semi-arid areas. Such deposits form mainly by the dissolution and precipitation of minerals (usually calcium carbonate). Caliche is a Spanish term and is broadly synonymous with hardpan, calcrete, or duricrust.

Cataclastic: a cataclastic rock is one produced by brittle fracturing and grinding within faults at relatively low temperatures ($<250°C$), therefore at high levels in the crust. At deeper levels, temperatures are high enough to deform minerals in a plastic, ductile manner to produce mylonites, the other major class of fault rocks.

Clasts: sedimentary clastic rocks, such as sandstones, conglomerates and breccias, are made up of fragments or "clasts" derived from the erosion of pre-existing rocks.

Culm facies: poorly defined term for mostly marine, dark sandstones and shales (with minor limestones and coals) of Carboniferous age. Used by Murchison and Sedgwick to characterize the Carboniferous rocks of SW England, and later extended to mainland Europe.

Detachment surface: a general term essentially synonymous with an extensional fault plane across which prominent movements of rock masses take place. Commonly used to describe well-defined, low-angle faults that cut cleanly across preexisting structures and penetrate to mid-crustal levels.

Diagenetic: change occurring within a sediment after deposition, typically during compaction, and aided by fluid movement.

Dip slope: when sedimentary strata are tilted by tectonic movements erosion commonly follows the angle (dip) of the bedding to produce a topographic dip slope. Compare with **scarp slope**.

Discordant: cutting across a pre-existing rock structure. For example an igneous dyke intrusion cutting discordantly across bedding or cleavage already present in the rocks being intruded.

Downlapping: downlap is a stratigraphic term used to describe the progressive deposition of inclined sediments out over a relatively flat surface, like sand at the front of a dune.

Euhedral: mineral crystals with sharp, well-formed crystal faces are referred to as having euhedral form.

Eustatic: an eustatic change in sea level is one that affects the oceans worldwide, an example being that resulting from ice-cap melting.

F_1: in areas of complex structural geology where rocks have been folded more than once, each successive folding event is given a number. Thus F_1 folds form during the first fold event, F_2 during the second, and so on.

Facies: see **Greenschist facies**.

Felsic: a rock rich in feldspar and quartz (a form of silica).

Flexural uplift: the Earth's lithosphere "floats" on the underlying asthenosphere rather like an iceberg. If more ice is added to an iceberg, it will float lower in the water, whereas if ice is removed the mass of the iceberg will rise. Similarly, when the lithosphere is thickened by mountain building, although mountains will rise in the upper part, the lithospheric mass overall will move downwards. Conversely, when a huge load is removed at the Earth's surface, such as by melting ice sheets or by thinning of the crust by erosion or extensional faults, there will be a compensatory uplift effect (isostatic rebound) and the lithosphere will slowly bend upwards, a process known as flexural uplift.

Foliated: see **anastomosing spaced foliation**.

Footwall: the rock beneath an inclined fault plane. See **hanging wall**.

Foreland basin: a depression that develops adjacent and parallel to a mountain belt. Thus the Ebro Basin is the southern foreland basin to the Pyrenees. Foreland basins form because the thickening lithospheric mass of the mountains causes the adjacent lithosphere to bend by a process known as lithospheric flexure.

Fracture zone: zone of brittle faulting within the upper crust.

Granitoid: a general term that denotes a granite-like igneous pluton rich in quartz and feldspar but avoids using the term "granite" which has a very specific definition based on its geochemical composition. Thus both granodiorites and granites are granitoid rocks.

Granodiorite: a **felsic pluton**ic rock similar to granite but with more plagioclase feldspar than alkali feldspar. The mafic minerals in granodiorites are usually biotite and/or hornblende. The continental crust has an average composition equivalent to that of a granodiorite.

Greenschist facies: deep within the Earth's crust, for any given rock composition, pressure and temperature, the group of new minerals that grows in the solid state during metamorphism is always the same. This fundamental truth has given rise to the concept of metamorphic facies. Each facies represents a particular range of pressure and temperature conditions and is therefore associated with minerals that are typical of that facies, depending on rock composition. The greenschist facies broadly represents rather low temperatures of 300–450°C and 2–6 kbars and is associated with minerals such as chlorite and muscovite which are common in slates and low-grade schists.

Grossular-andradites: garnet minerals. There are many different types of garnet minerals. This is because, while they are all silicates, the amounts of magnesium, iron, manganese, calcium, aluminium and other elements substituting in their crystal structure is highly variable and depends on the original rock composition. Grossular garnet is rich in calcium and aluminium whereas andradite garnet contains more ferric iron and titanium in place of aluminium. A "grossular-andradite" is a chemical mixture of the two types, both of which are typical **calc-silicate** minerals.

Glossary

Half-graben: a graben is a symmetrical rift valley depression dropped down on both sides by normal faults. Where there is only one faulted margin the resulting asymmetrical structure is known as a half-graben.

Hanging wall: the rock above an inclined fault plane. See **footwall**.

Hornfels: hornfels is a rock baked by the heating (contact-metamorphic) effects of a nearby igneous intrusion. The lack of directed pressure during heating means that new minerals are not oriented and therefore do not define a metamorphic **foliation** but instead form a tight, interlocking network which makes the rock hard and splintery. Unimpeded by directed pressure, new minerals such as cordierite commonly form rounded crystals in a finer matrix which imparts a spotty texture to the rock (thus "spotted slate").

Horst-and-graben: a horst is a raised block of land forming high ground between fault-rift valleys (graben). See **half-graben**.

Imbrication: repeated slicing of rocks by multiple faults.

Isostatically: in Earth Science isostasy refers to the gravitational equilibrium between the lithosphere and underlying asthenosphere such that the tectonic plates "float" at an elevation that depends on their density and thickness. See **flexural uplift**.

L_1: in areas of complex structural geology where rocks have been deformed more than once, each successive deformation event is given a number. Any linear structures (**lineation**) produced during deformation, such as a fold axis, will be appended a number accordingly. Thus the fold axis of an F_1 fold is an L_1 structure.

Lineation: in structural geology, features that can be defined as a line, such as a fold axis, or the intersection of cleavage on a bedding plane, are referred to as lineations.

Lithification: the process by which sediments harden and become more compact (more lithified) as their porosity decreases and they lose fluids under the increasing weight of overlying rock.

Lithostratigraphic correlations: Lithostratigraphy involves the study of rock layers, ideally defining detailed units with the aid of fossils, radiometric ages, and distinctive lithological sequences. Such units will be traceable on a map and may be recognizable elsewhere, so that different areas can be correlated. Unfortunately, in metamorphic areas such as the Collserola Hills the lack of information regarding age, and the overprint of folding, cleavage formation, and contact metamorphism impede such correlation.

Loess: homogeneous aolian sediment formed by the accumulation of windblown silt-sized mineral particles.

Ma: abbreviation of *megaannum* and denoting millions of years.

Matrix: where larger crystals or sedimentary clasts lie embedded in finer material, the latter is referred to as a matrix.

Metabasites: iron-magnesium-rich, silica-poor igneous rocks (such as basalts) which have been recrystallised in the solid state during metamorphism. Common types of metabasites include greenschists and amphibolites.

Metacalcareous: a metamorphosed rock rich in calcium and so containing re-crystallised calcite or other calcium minerals.

Metalimestones: a limestone which has been recrystallised in the solid state during the application of heat and pressure within the Earth (metamorphism).

Metamorphic aureole: the transfer of heat from an igneous intrusion into the adjacent rock envelope ("country rock") promotes the growth of new minerals. The greatest effect of this "contact metamorphism" will be seen in the country rock closest to the contact with the intrusion. The final result is commonly recorded as broadly concentric zones of metamorphism, decreasing in intensity away from the intrusion and referred to as a metamorphic aureole.

Metamorphic grade: metamorphic rocks are referred to as having lower or higher grade depending on the levels of heating and pressure to which they have been subjected. See **greenschist facies**.

Metapelites: metamorphosed argillaceous rocks such as mudstone.

Metapsammites: metamorphosed arenaceous rocks such as sandstone.

Metaquartzites: a metamorphosed quartzite (pure, quartz-rich sandstone).

Metasediments: metamorphosed sedimentary rocks.

Monoclinal folds: an assymmetric fold with only one side or limb. For example a sudden change of dip from gentle to steep then back to gentle as before produces a monoclinal fold with just one (steeply dipping) limb.

Palaeosols: fossilized ancient soil deposits.

Phenocrystic: a phenocryst is a large crystal in an otherwise finer-grained igneous rock. Such a texture implies early nucleation and growth of the larger mineral, before the surrounding magma froze around it. See **porphyritic**.

Pluton: a body of igneous rock within the Earth that has intruded as magma then solidified slowly enough to allow growth of mineral crystals large enough to be easily seen with the naked eye.

Porphyritic: a porphyritic texture in igneous rocks describes large mineral phenocrysts embedded in a finer matrix.

Porphyroblastic: a porphyroblast is a large mineral crystal (such as garnet) which has grown in a metamorphic rock. Unlike phenocrysts, which grow in magma, a porphyroblast grows in solid rock.

Protoliths: metamorphic rocks are formed as a result of the application of heat and pressure on previously existing rock. The original rock that existed prior to its metamorphism is known as the protolith.

PT conditions: in metamorphic geology it is important to ascertain the pressure (P) and temperature (T) represented by the mineral assemblage in the rock under study. Such PT conditions will reveal much about the history of the rock and the regional geological setting under which it formed.

Radiometric ages: measurement in a rock or mineral of a radioactive isotope and its decay products, given a known decay rate, provides useful information relating to the absolute (or "radiometric") age of that rock or mineral. The method

Glossary

relies on there being no loss of the daughter decay product, in other words a "closed system" must exist. In the case of plutonic intrusions such as the Barcelona granodiorite, application of this technique to different minerals can reveal a range of ages each of which represents a time when the mineral cooled sufficiently to become a closed system, entrapping subsequent release of radioactive decay products. Such information records geological moments in the crystallization and slow cooling history of the intrusion.

Recumbent fold: a fold with a flat axial plane. See **axial planar fabric**.

Rift basins: a linear, downfaulted area bounded by extensional faults and typically receiving sediment eroded from higher ground. See **half-graben**.

S_1 foliation: foliation in a metamorphic rock is a planar structure (or "surface") such as cleavage produced by the alignment of new minerals growing under directed pressure. The first metamorphic foliation to be formed in a rock is referred to as S_1, having been produced by the first deformation event (D_1). It is common for foliations to be folded by later deformation events (D_2, D_3, etc.) so, for example, an S_1 foliation may be folded by F_2 folds during D_2, and so on. If the original bedding is still discernable in a metamorphic rock it is referred to as S_0.

Scarp slope: the opposite of a **dip slope**, where erosion cuts steeply down across the bedding. A classic dip-and-scarp topographic profile, such as that displayed by the hill of Montjuïc in Barcelona, has a gentle dip slope and a steep scarp slope.

Stoping: a mechanism aiding the intrusion of hot **granitoid** magmatic bodies into the surrounding country rock, the heating of which induces thermal expansion and fracturing. The granitoid magma then passively eases into the fractures. See **xenoliths**.

Synorogenic: Orogeny involves a long period when tectonic plates interact to produce a belt of severe deformation of the Earth's crust. Any geological event that takes place during the orogeny, such as the deposition of sediment in a basin adjacent to a growing mountain belt, is referred to synorogenic.

Terrane: since the 1980s geological usage of the word terrane has focused on describing dislocated slices of crust ("tectonostratigraphic terranes") moved to a new position along major faults. The concept has been most useful where transcurrent faults have moved these slices long distances horizontally along a continental margin, with each slice showing a distinctive stratigraphy and tectonic history that is quite different from now-adjacent areas against which they have become positioned or "accreted". Along many continental plate margins, a long history of oblique plate movement, where plates graze past each other rather than directly collide, has produced zones of complex geology comprising many of these terranes (a "continental terrane collage").

Transpressional inversion: geological use of inversion simply means an originally low-lying area that has later been uplifted by faulting (note that in this sense the term has nothing to do with overturning strata). Inversion is achieved either by simple compression (producing reverse faults such as thrusts), or by some combination of compression and transcurrent (strike-slip) fault movement referred to in brief as transpression.

Xenoliths: in igneous geology a xenolith ("foreign stone") is a piece of country rock which has become incorporated within an intruding magma. In the case of the Barcelona granodiorite, pieces of the metamorphic country rock have been enveloped by magma during **stoping**.

References

REFERENCES AND FURTHER READING

Ábalos, B., Carreras, J., Druguet, E., Escuder Viruete, J., Gómez Pugnaire, M.T., Lorenzo Alvarez, S., Quesada, C., Rodríguez Fernández, L.R. & Gil Ibarguchi, J.I. 2002. Variscan and Pre-Variscan Tectonics. In: Gibbons, W. & Moreno, T. (eds) *The Geology of Spain*. The Geological Society of London, London, 155–183.

Alborch, J., Civis, J. & Martinell, J. 1980. Nuevas aportaciones micropaleontológicas al conocimiento del Neógeno del Baix Llobregat (Barcelona). *Acta Geológica Hispánica*, **15**, 85–90.

Almera, J. 1900. *Mapa geológico y topográfico de la provincia de Barcelona. Región primera o de contornos de la capital*. Escala 1:40 000. 2ª edición.

Amblàs, D., Canals, M., Urgeles, R., Lastrasa, G., Piquete, C., Hughes-Clarke, J.E., Casamor, J.L. & Calafat, A.M. 2006. Morphogenetic mesoscale analysis of the northeastern Iberian margin, NW Mediterranean Basin. *Marine Geology*, **234**, 3–20.

Anadón, P., Colombo, F., Esteban, M., Marzo, M., Robles, S., Santanach, P. & Solé Sugrañés, L.L. 1979. Evolución tectonoestratigráfica de los Catalánides. *Acta Geológica Hispánica*, **14**, 242–270.

Anadón, P., Julivert, M. & Sáez, A. 1983. El Carbonifero de las Cadenas Costeras Catalanas. In: Martínez-Díaz, C. (ed.) *Carbonifero y Permico de España*. IGME, Madrid, 331–336.

Anadón, P. & Roca, E. 1996. Geological setting of the Tertiary basins of Northeastern Spain. In: Friend, P.F. & Dabrio, C.J. (eds) *Tertiary Basins of Spain: The Stratigraphic Record of Crustal Kinematics*. Cambridge University Press, Cambridge, 68–76.

Arranz Herrero, M. 2003. *La Rambla de Barcelona. Estudi d'història urbana*. Rafael Dalmau, Barcelona 128 pp.

Aurell, M., Meléndez, G., Olóriz, F., Bádenas, B., Caracuel, J.E., García-Ramos, J.C., Goy, A., Linares, A., Quesada, S., Robles, S., Rodríguez-Tovar, F.J., Rosales, I., Sandoval, J., Suárez de Centi, C., Tavera, J.M. & Valenzuela, M. 2002. Jurassic. In: Gibbons, W. & Moreno, T. (eds) *The Geology of Spain*. The Geological Society of London, London, 213–253.

Banda, E. 1996. Deep crustal expression of Tertiary basins in Spain. In: Friend, P.F. & Dabrio, C.J. (eds) *Tertiary Basins of Spain: The Stratigraphic Record of Crustal Kinematics*. Cambridge University Press, Cambridge, 15–19.

Banda, E. & Correig, A. 1984. The Catalan earthquake of February 2 1428. *Engineering Seismology*, **20**, 89–97.

Banda, E. & Santanach, P. 1992. The Valencia trough (western Mediterranean): an overview. *Tectonophysics*, **208**, 183–202.

Barberá, X., Cabrera, L., Marzo, M., Parés, J.M. & Agustí, J. 2001. A complete terrestrial Oligocene magnetostratigraphy from the Ebro Basin, Spain. *Earth and Planetary Science Letters*, **187**, 1–16.

References

Barrois, Ch. 1893. Observaciones sobre el terreno Silúrico de los alrededores de Barcelona. *Boletín de la Comisión del Mapa Geológico de España*, **19**, 245–260.

Barrois, Ch. 1901. Note sur les Graptolites de la Catalogne et leurs relations avec les étages graptolitiques de la France. *Bulletin de la Societe Geologique de la France*, **1**, 637–648.

Bartrina, M.T., Cabrera, L., Jurado, M.J., Guimerà, J. & Roca, E. 1992. Evolution of the central Catalan margin of the Valencia trough (western Mediterranean). *Tectonophysics*, **203**, 219–247.

Batlle, A. 1976. Influència dels paràmetres geològics a l'excavació del Túnel del Turó de la Rovira (Barcelona). *Revista del Instituto de Investigaciones Geológicas de la Diputación Provincial (Universidad de Barcelona)*, **31**, 55–72.

Bonneville, J.N. 1981. Les Cupae de Barcelone: les origines du type monumental. *Melanges de la Casa Velázquez*, **17**, 5–38.

Cabrera, L. (ed.) 1994. El margin continental catalán (I): El marco de la Cuenca Catalana-Balear. *Acta Geológica Hispánica*, **29**, 1–87.

Cabrera, L. & Calvet, F. 1996. Onshore Neogene record in NE Spain: Vallès-Penedès and El Camp half-grabens (NW Mediterranean). In: Friend, P.F. & Dabrio, C.J. (eds) *Tertiary Basins of Spain: The Stratigraphic Record of Crustal Kinematics*. Cambridge University Press, Cambridge, 97–105.

Cabrera, L. & Santanach, P. 1979. Precisions sobre la disposició estructural dels terrenys triàsics de Vallcarca (Barcelona). *Butlletí Institució Catalana d'Història Natural*, **43**, 73–77.

Castro, A., Corretgé, L.G., De La Rosa, J., Enrique, P., Martínez, F.J., Pascual, E., Lago, M., Arranz, E., Galé, C., Fernández, C., Donaire, T. & López, S. 2002. Palaeozoic Magmatism. In: Gibbons, W. & Moreno, T. (eds) *The Geology of Spain*. The Geological Society of London, London, 117–153.

Checa, A., Díaz, J.I., Farrán, M. & Maldonado, A. 1988. Sistemas deltaicos holocenos de los ríos Llobregat, Besós y Foix: modelos evolutivos transgresivos. *Acta Geológica Hispánica*, **23**, 241–255.

Clavell, E. & Berastegui, X. 1991. Petroleum geology of the Gulf of València. In: Spencer, A. M. (ed.) *Generation, accumulation, and production of Europe's hydrocarbons*. Special Publication of the European Association of Petroleum Geoscientists, **1**. Oxford University Press, Oxford, 355–368.

Colmenero, J.R., Fernández, L.P., Moreno, C., Bahamonde, J.R., Barba, P., Heredia, N. & González, F. 2002. Carboniferous. In: Gibbons, W. & Moreno, T. (eds) *The Geology of Spain*. The Geological Society of London, London, 93–116.

Conillera, P. 1991. *Descobrir el medi urbà. 8. L'aigua de Montcada. L'abastamanet municipal d'aigua de Barcelona. Mil Anys d'Història*. Institut d'Ecologia Urbana de Barcelona. Ajuntament de Barcelona.

Dañobeitia, J.J., Alonso, B. & Maldonado, A. 1990. Geological framework of the Ebro continental margin and surrounding areas. *Marine Geology*, **95**, 265–287.

References

Depape, G. & Solé Sabaris, L. 1934. Constitució geològica del Turó de Montgat. *Butlletí Institució Catalana d'Història Natural*, **34**, 138–148.

Díaz, J.I., Nelson, C.H., Barber Jr., J.H. & Giró, S. 1990. Late Pleistocene and Holocene sedimentary facies on the Ebro continental shelf. *Marine Geology*, **95**, 333–352.

Elías, J. 1931. Esfondraments a Montcada i a Martorell en començar el període pliocènic. *Butlletí de la Institució Catalana d'Història Natural*, **30**, 60–65.

Enrique, P. 1979. Las rocas graníticas de la Cordillera Litoral Catalana entre Mataró y Barcelona. *Acta Geológica Hispánica*, **13**, 81–86.

Ercilla, G., Farrán, M., Alonso, B. & Díaz, J.I. 1994. Pleistocene progradational growth pattern of the northern Catalonia continental shelf (northwestern Mediterranean). *Geo-Marine Letters*, **14**, 264–271.

Figueras, S., Villaseñor, A., Frontera, T., Olivera, C., Fleta, J., Ruiz, M., Díaz, J., Gallart, J. & Vergès, J. 2006. Analysis of the September 2004 seismic crisis in the area of the 1428 earthquake (I0=IX), Eastern Pyrenees (Spain). *First European Conference on Earthquake Engineering and Seismology, Geneva*, Paper Number **1034**, 1–8.

Fontboté, J.M., Guimerà, J., Roca, E., Sàbat, F., Santanach, P. & Fernández-Ortigosa, F. 1990. The Cenozoic geodynamic evolution of the Valencia trough (western Mediterranean). *Revista de la Sociedad Geológica de España*, **3**, 249–259.

Friend, P.F. & Dabrio, C.J. (eds) 1996. *Tertiary Basins of Spain: The Stratigraphic Record of Crustal Kinematics*. Cambridge University Press, Cambridge, 400 pp.

Gámez, D., Simó, J.A., Vázquez-Suñé, E., Salvany, J.M. & Carrera, J. 2005. Variación de las tasas de sedimentación en el Complejo Detrítico Superior del Delta del Llobregat (Barcelona): su relación con causas eustáticas, climáticas y antrópicas. *Geogaceta*, **38**, 175–179.

Gámez, D., Simó, J.A., Lobo, F.J., Barnolas, A., Carrera, J. & Vázquez-Suñé, E. 2009. Onshore-offshore correlation of the Llobregat deltaic system, Spain: Development of deltaic geometries under different relative sea-level and growth fault influences. *Sedimentary Geology*, **217**, 65–84.

García-Alcalde, J.L., Carls, P., Pardo Alonso, M.V., Sanz López, J., Soto, F., Truyols-Massoni, M. & Valenzuela-Ríos, J.I. 2002. Devonian. In: Gibbons, W. & Moreno, T. (eds) *The Geology of Spain*. The Geological Society of London, London, 67–91.

García-López, S., Julivert, M., Soldevila, J., Truyols-Massoni, M. & Zammarreño, I. 1990. Biostratigrafía y Facies de la sucesión carbonatada del Silúrico superior y Devónico Inferior de Santa Creu d'Olorda (Cadenas Costeras Catalanas, EN de España). *Acta Geológica Hispánica*, **25**, 141–168.

Gaspar-Escribano, J.M., Van Wees, J.D., ter Voorde, M., Cloetingh, S., Roca, E., Cabrera, L., Muñoz, J.A., Ziegler, P.A. & García-Castellanos, D. 2001. 3D Flexural Modeling of the Ebro Basin (NE Iberia). *Geophysical Journal International*, **145**, 349–367.

References

Gaspar-Escribano, J.M., ter Voorde, M., Roca, E. & Cloetingh, S. 2002. Mechanical (de)-coupling of the lithosphere in the Valencia Trough (NW Mediterranean): what does it mean? *Earth and Planetary Science Letters*, **210**, 291–303.

Gaspar-Escribano, J.M., Garcia-Castellanos, D., Roca, E. & Cloetingh, S. 2004. Cenozoic vertical motions of the Catalan Coastal Ranges (NE Spain): The role of tectonics, isostasy, and surface transport. *Tectonics*, **23**, TC1004, doi:10.1029/2003TC001511.

Gibbons, W. & Moreno, T. (eds) 2002. *The Geology of Spain*. The Geological Society of London, London, 632 pp. ISBN: 978-1-86239-127-7.

Gil Ibarguchi, J.I. & Julivert, M. 1988. Petrología de la aureola metamórfica de la granodiorita de Barcelona en la Sierra de Collcerola (Tibidabo). *Estudios Geológicos*, **44**, 353–374.

Gil Ibarguchi, J.I., Navidad, M. & Ortega, L.A. 1990. Ordovician and Silurian igneous rocks and orthogneisses in the Catalonian Coastal Ranges. *Acta Geológica Hispánica*, **25**, 23–29.

Gillet, S. & Vicente, J. 1961. Nuevo yacimiento pliocénico de facies salobres en el subsuelo de Barcelona al este del Tibidabo. *Notas y comunicaciones del Instituto Geológico y Minero de España*, **63**, 253–292.

Gómez-Gras, D., Parcerisa, D., Calvet, F., Porta, J., Solé de Porta, N. & Civís, J. 2001. Stratigraphy and petrology of the Miocene Montjuïc delta (Barcelona, Spain). *Acta Geológica Hispánica*, **36**, 115–136.

Gutiérrez-Marco, J.C., Robardet, M., Rábano, I., Sarmiento, G.N., San José Lancha, M.A., Herranz Araújo, P.H. & Pieren Pidal, A.P. 2002. Ordovician. In: Gibbons, W. & Moreno, T. (eds) *The Geology of Spain*. The Geological Society of London, London, 31–49.

Hughes, R. 1992. *Barcelona*. Harvill Press, London, 573 pp.

Institut Cartogràfic de Catalunya. 1999. *Atles sísmic de Catalunya*. Volume **1**: seismicity catalogue.

Institut Cartogràfic de Catalunya. 2000. *Mapa geotècnic de Barcelona 1:25 000*.

Institut Cartogràfic de Catalunya. 2002. *Mapa geològic de Catalunya* (2nd edition) 1: 250 000.

Institut Cartogràfic de Catalunya. 2005. *Barcelonés: Mapa geològic comarcal de Catalunya 1:50 000*.

Institut Geològic de Catalunya/Institut Cartogràfic de Catalunya. 2009a. *Mapa geològic de les zones urbanes: Barcelona-Horta 1:5 000*.

Institut Geològic de Catalunya/Institut Cartogràfic de Catalunya. 2009b. *Mapa geològic de les zones urbanes: Turons de Barcelona 1:5 000*.

Jalut, G., Amat, A.E., Bonnet, L., Gauquelin, T. & Fontugne, M. 2000. Holocene climate changes in the Western Mediterranean, from south-east France to south-east Spain. *Palaeogeography, Palaeoclimatology, Palaeoecology*, **160**, 255–290.

References

Jiménez, J., Suriñach, E., Fleta, J. & Goula, X. 1996. Recent vertical movements from high-precision leveling data in northeast Spain. *Tectonophysics*, **263**, 149–161.

Julià, R. 1977. Características litológicas de las "rieres" del Pla de Barcelona. *Cuadernos de Arqueología e Historia de la ciudad*, **17**, 21–26.

Julivert, M. & Durán, H. 1990a. Paleozoic stratigraphy of the Central and Northern part of the Catalonian Coastal Ranges (NE Spain). *Acta Geológica Hispánica*, **25**, 3–12.

Julivert, M. & Durán, H. 1990b. The Hercynian structure of the Catalonian Coastal Ranges (NE Spain). *Acta Geológica Hispánica*, **25**, 13–21.

Julivert, M., Durán, H., Richards, R.B. & Chapman, A.J. 1985. Siluro–Devonian graptolite stratigraphy of the Catalonian Coastal Ranges. *Acta Geológica Hispánica*, **20**, 199–207.

Lafuerza, S., Canals, M., Casamore, J.L. & Devincenzi, J.M. 2005. Characterisation of deltaic sediment bodies based on in situ CPT/CPTU profiles: A case study on the Llobregat delta plain, Barcelona, Spain. *Marine Geology*, **222–223**, 497–510.

Liñán, E., Gozalo, R., Palacios, T., Gámez Vintaned, J.A., Ugidos, J.M. & Mayoral, E. 2002. Cambrian. In: Gibbons, W. & Moreno, T. (eds) *The Geology of Spain*. The Geological Society of London, London, 17–29.

Liquete, C., Canals, M., Lastras, G., Amblas, D., Urgeles, R., De Mol, B., De Batist, M. & Hughes-Clarke, J.E. 2007. Long-term development and current status of the Barcelona continental shelf: A source-to-sink approach. *Continental Shelf Research*, **27**, 1779–1800.

Llopis, N. 1942. *Los terrenos cuaternarios del Llano de Barcelona*. Publicaciones del Instituto Geológico y Topográfico de la Diputación de Barcelona, Barcelona, 51 pp.

Llopis, N., Vía Boada, L. & De Villalta, J. F. 1969. Sobre el límite Silúrico-Devónico en Santa Creu d'Olorde. *Cuadernos de Geología Ibérica*, **1**, 3–20.

London, D. 2009. The origin of primary textures in granitic pegmatites. *Canadian Mineralogist*, **47**, 697–724.

López-Blanco, M. 2002. Sedimentary response to thrusting and fold growing on the SE margin of the Ebro basin (Paleogene, NE Spain). *Sedimentary Geology*, **146**, 133–154.

López-Blanco, M., Marzo, M., Burbano, D.W., Vergés, J., Roca, E., Anadón, P. & Piña, J. 2000. Tectonic and climatic controls on the development of foreland fan deltas: Montserrat and Sant Llorenç del Munt systems (Middle Eocene, Ebro basin, NE Spain). *Sedimentary Geology*, **138**, 17–39.

López-Gómez, J., Arche, A. & Pérez-López, A. 2002. Permian and Triassic. In: Gibbons, W. & Moreno, T. (eds) *The Geology of Spain*. The Geological Society of London, London, 185–212.

References

Lunar, R., Moreno, T., Lombardero, M., Regueiro, M., López-Vera, F., Martínez del Olmo, W., Mallo García, J.M., Saenz de Santa Maria, J.A., García-Palomero, F., Higueras, P., Ortega, L. & Capote, R. 2002. Economic and environmental geology. In: Gibbons, W. & Moreno, T. (eds) *The Geology of Spain*. The Geological Society of London, London, 473–510.

Lyell, C. 1830. *Principles of geology (volume 1), being an attempt to explain the former changes of the Earth's surface, by reference to causes now in operation.* Murray, London, 511 pp.

Lyell, C. 1832. *Principles of geology (volume 2), being an attempt to explain the former changes of the Earth's surface, by reference to causes now in operation.* Murray, London, 330 pp.

Lyell, C. 1833. *Principles of geology (volume 3), being an attempt to explain the former changes of the Earth's surface, by reference to causes now in operation.* Murray, London, 558 pp.

Maillard, A. & Maufrett, A. 1999. Crustal structure and riftogenesis of the Valencia Trough (north-western Mediterranean Sea). *Basin Research*, **11**, 357–379.

Marcet, J. 1933. Les formacions paleozoiques dels encontorns del Papiol. *Memorias de la Real Academia de Ciencias y Artes de Barcelona*, **23**, 189–202.

Marcet, J. 1960. Las formaciones paleozoicas de los alrededores de Santa Cruz de Olorde. *Notas y Comunicaciones del IGME*, **57**, 135–163.

Marqués, M.A. 1966. Observaciones sobre el Cuaternario del delta del Llobregat (Barcelona). *Acta Geológica Hispánica*, **1**, 9–12.

Marqués, M.A. 1975. Las formaciones cuaternarias del Delta del Llobregat. *Acta Geológica Hispánica*, **10**, 21–28.

Martín-Chivelet, J., Berástegui, X., Rosales, I., Vilas, L., Vera, J.A., Caus, E., Gräfe, K., Mas, R., Puig, C., Segura, M., Robles, S., Floquet, M., Quesada, S., Ruiz-Ortiz, P.A., Fregenal-Martínez, A.A., Salas, R., Arias, C., García, A., Martín-Algarra, A., Meléndez, M.N., Chacón, B., Molina, J.M., Sanz, J.L., Castro, J.M., García-Hernández, M., Carenas, B., García-Hidalgo, J., Gil, J. & Ortega, F. 2002. Cretaceous. In: Gibbons, W. & Moreno, T. (eds) *The Geology of Spain*. The Geological Society of London, London, 255–292.

Martínez del Olmo, W. 1996. Depositional sequences in the Gulf of Valencia Tertiary basin. In: Friend, P.F. & Dabrio, C.J. (eds) *Tertiary Basins of Spain: The Stratigraphic Record of Crustal Kinematics*. Cambridge University Press, Cambridge, 55–67.

Masana, E. 1996. Neotectonic features of the Catalan Coastal Ranges, Northeastern Spain. *Acta Geológica Hispánica*, **29**, 107–121.

Murcia, J. 2008. *Seismic analysis of Santa Maria del Mar church in Barcelona.* MSc thesis, Technological University of Catalonia (UPC) 113 pp.

Olivera, C., Banda, E. & Roca, A. 1991. An outline of historical seismicity studies in Catalonia. *Tectonophysics*, **193**, 231–235.

References

Olivera, C., Susagna, T., Roca, A. & Goula, X. 1992. Seismicity of the Valencia Trough and surrounding areas. *Tectonophysics*, **203**, 99–109.

Olivera, C., Redondo, E., Lambert, J., Riera, A. & Roca, A. 2006. *Els terratrèmols dels segles XIV i XV a Catalunya*. Monografies de l'Institut Cartogràfic de Catalunya, **30**. Barcelona 407 pp.

Plusquellec, Y., Fernández-Martínez, E., Sanz López, J., Soto, F., Magrans, J. & Ferrer, E. 2006. Tabulate corals and stratigraphy of Lower Devonian and Mississippian rocks near Barcelona (Catalonian Coastal Ranges), northeast Spain. *Revista Española de Paleontología*, **22**, 175–192.

Riba i Arderiu, O. & Colombo i Piñol, F. 2009. *Barcelona: la ciutat Vella i el Poblenou, assaig de geologia urbana*. Institut d'Estudis Catalans i Reial Academia de Ciències i Arts de Barcelona, Barcelona, 278 pp.

Robardet, M. & Gutiérrez-Marco, J.C. 2002. Silurian. In: Gibbons, W. & Moreno, T. (eds.) *The Geology of Spain*. The Geological Society of London, London, 51–66.

Roca, E. & Desegaulx, P. 1992. Analysis of the geological evolution and vertical movements in the Valencia Trough area, western Mediterranean. *Marine and Petroleum Geology*, **9**, 167–185.

Roca, E. & Guimerà, J. 1992. The Neogene structure of the eastern Iberian margin: structural constraints on the crustal evolution of the Valencia Trough (western Mediterranean). *Tectonophysics*, **203**, 203–218.

Roca, E., Sans, M., Cabrera, L. & Marzo, M. 1999. Oligocene to Middle Miocene evolution of the central Catalan margin (northwestern Mediterranean). *Tectonophysics*, **315**, 209–233.

Rubio, C. & Kindelán, A. 1910. Hidrología subterránea del llano de Barcelona. *Boletín Comité Mapa Geológico de España*, **30**, 93–105.

Sàbat, F., Roca, E., Muñoz, J.A., Vergés, J., Santanach, P., Sans, M., Masana, E., Estévez, A. & Santisteban, C. 1995. Role of extensión and compression in the evolution of the eastern margin of Iberia: the ESCI-València Trough seismic profile. *Revista de la Sociedad Geológica de España*, **8**, 431–448.

Salas, R. & Casas, A. 1993. Mesozoic extensional tectonics, stratigraphy and crustal evolution during the Alpine cycle of the eastern Iberian basin. *Tectonophysics*, **228**, 33–55.

Salicrú i Lluch, R. 1995. The 1448 earthquake in Catalonia. Some effects and local reactions. *Annali di Geofisica*, **38**, 503–513.

San Miguel, M. 1929. Las pizarras cristalinas de silicato cálcico de la zona metamórfica del Tibidabo. *Memorias de la Real Academia de Ciencias y Artes de Barcelona*, **21**, 513–530.

Sanz Parera, M. 1988. *El Pla de Barcelona: constitució i característiques físiques*. Els Llibres de la Frontera, Serie Coneguem Catalunya 25, Amelia Romero, Sant Cugat, 144 pp.

References

Sebastian, A., Reche, J. & Duran, H. 1990. Hercynian metamorphism in the Catalonian Coastal Ranges. *Acta Geológica Hispánica*, **25**, 31–38.

Simó, J.A., Gàmez, D., Salvany, J.M., Vázquez-Suñé, E., Carrera, J., Barnolas, A. & Alcalà, F.J. 2005. Arquitectura de facies de lo deltas cuaternarios del río Llobregat, Barcelona, España. *Geogaceta*, **38**, 171–174.

Solé, J., Cosca, M., Sharp, Z. & Enrique, P. 2002. $^{40}Ar/^{39}Ar$ Geochronology and stable isotope geochemistry of Late Hercynian intrusions from north-eastern Iberia with implications for argon loss in K-feldspar. *International Journal of Earth Sciences*, **91**, 865–881.

Solé de Porta, N., Calvet, F. & Torrento, L. 1987. Análisis palinológico del Triásico de los Catalánides (NE de España). *Cuadernos de Geología Ibérica*, **11**, 237–254.

Stallybrass, C.O. 1931. *The Principles of Epidemiology*. Macmillan, New York, 696 pp.

Torné, M. 1996. The lithosphere of the Valencia Trough: a brief review. In: Friend, P.F. & Dabrio, C.J. (eds) *Tertiary Basins of Spain: The Stratigraphic Record of Crustal Kinematics*. Cambridge University Press, Cambridge, 49–55.

Valenciano, A. & Sanz, F. 1967. Algunas consideraciones sobre la edad y paleogeografía de las gravas de Castellbisbal. *Acta Geológica Hispánica*, **2**, 55–59.

Valenciano, A. & Sanz, F. 1981. Nota sobre la estructura del Paleozoico y la presencia del Triásico en el noroeste de la ciudad de Barcelona (Vallcarca, El Coll, Monte Carmelo). *Revista del Instituto de Investigaciones Geológicas de la Diputación Provincial, Universidad de Barcelona*, **35**, 61–69.

Vázquez-Suñé, E., Abarca, E., Carrera, J., Capino, B., Gámez, D., Pool, M., Simó, T., Batlle, F., Niñerola, J.M. & Ibáñez, X. 2006. Groundwater modelling as a tool for the European Water Framework Directive (WD) application: The Llobregat case. *Physics and Chemistry of the Earth*, **31**, 1015–1029.

Vézian, A. 1856. *Du terrain post-pyrénéen des environs de Barcelona et de ses rapports avec les formations correspondantes du bassin de la mediterranée*. Thèse de Géologie, Université de Montpellier, 115 pp.

Vía Boada, L. & Padreny, J. 1972. Historia bibliográfica sobre geología de Montjuïc (Barcelona). *Revista del Instituto de Investigaciones Geológicas Diputación Provincial de Barcelona*, **27**, 5–63.

Villas, E., Durán, H. & Julivert, M. 1987. The Upper Ordovician Clastic Sequence of the Catalonian Coastal Ranges and its Brachiopod Fauna. *Neues Jahrbuch für Geologie und Paläontologie - Abhandlungen*, **174**, 55–74.

Virgili, C. 2007a. Lyell and the Spanish Geology. *Geologica Acta*, **5**, 119–126.

Virgili, C. 2007b. Charles Lyell and scientific thinking in geology. *Comptes Rendus Geoscience*, **339**, 572–584.

Yáñez, A. 1822. Ensayo de la descripción mineralógica de la montaña de Montjuïc. *Periódico de la Sociedad de la Salud Pública de Cataluña (Barcelona)*, **1**, 47–58 and 142–150.